AF540865

GLOBALISATION AND SUSTAINABLE AGRICULTURE IN INDIA

GLOBALISATION AND SUSTAINABLE AGRICULTURE IN INDIA

Edited by
Rajib Lochan Panigrahy
&
Dr. L.R. Patro

DISCOVERY PUBLISHING HOUSE
NEW DELHI-110002

First Published-2007

ISBN 978-81-8356-198-3

Published by

DISCOVERY PUBLISHING HOUSE
4831/24, Ansari Road, Prahlad Street,
Darya Ganj, New Delhi-110002 (India)
Phone: 23279245 • Fax: 91-11-23253475
E-mail: dphbooks@rediffmail.com
dphtemp@indiatimes.com

Printed at:
Sachin Printers, Delhi

Editorial

Agriculture is harbinger of our Indian society and main input and output to the earning. Majority of Indian citizen lives on agriculture as their main livelihood. The other sectors of economy depend on agriculture as the main sector and factor of production. In the pace of time, development of science made the sector costlier, environmental degradation and irregular monsoon fails crop and cropping pattern changes day-by-day. Due to improvement in communication system people migrating to urban-industrial areas for their livelihood, which earns more than agriculture, and give comfort to them. Now-a-days, the rural youth are thinking of prestige, less labour more earning, show themselves more lucrative on others. The modern method of cropping is more expenses than the traditional cropping. To cope with modern commercial cropping, credit is the source of investment. Due to less importance on agriculture produce for sale in different markets from domestic to international and agriculture produce are perishable in nature, destroyed/damaged by monsoon and natural calamities give more loss to the farmers which is the evidence of farmer's suicidal death. So most of the modern commercial crops are on paper, research in laboratories. Day-by-day temperature of air is going high, which creates adverse affects on the crop. In the nutshell, every Indian should think of environment for living of their self, which ultimately will protect agricultural produce.

EDITORS

Contents

CHAPTER 1

Development of Indian Agricultural Sector in Globalisation Era: A Management Look

Sudhansu Sekhar Nayak* and Anil Kumar Sahu**

Introduction

More than 58% of country's population depends on agriculture, a sector producing only 22% of GDP. The agriculture and allied sector witnessed a growth of 9.1% in 2003-04, which fell steeply to 1.1% in the current fiscal year. Favourable monsoon facilitated an impressive growth rate of 9.6% in 2003-04 on the back of negative growth in the preceding year. However, deficient rainfall from the southwest monsoon is estimated to have caused a significant decline in kharif crops production in the current year.

While looking at some of the agricultural products, one finds that India is the largest producer of tea, jute and jute like fibre. In is not only the largest producer but also largest consumer of tea in the world. India accounts for around 14% of the world trade in tea. Indian tea is exported in various forms such as bulk tea, packet tea, tea bags, instant tea etc., to more than 80 countries of the world. Among livestock cattle and buffalo are found maximum in India. Indian total milk production is highest in the world. India has also the privilege of having the 1st rank in total irrigated land in area terms in the world. Among cereals production, India is placed third, having second largest production in wheat and rice and the largest production in pulses. However, the full potential of

*Lecturer in Commerce, R. N. College, Dura, Berhampur, Ganjam (Orissa).

**Reader in MBA, Berhampur University, Bhanja Bihar (Orissa).

Indian agriculture as a profitable activity hasn't been realized yet. Agriculture upliftment will not only benefit farmers and a large section of the rural poor, but also will give fillip to overall growth of the economy through the backward and forward linkages of agriculture with the rest of the economy.

Priority must be given to livestock's and fisheries, horticulture, organic farming, commercial crops and agro-processing, as these are the potential areas of high growth. Further, rationalization of minimum support price regime and introduction of other risk-mitigation measures, improvements in rural infrastructure are essential for sustaining high agricultural growth. It is conceived that reforms in legislations, strengthening R&D and improvements in post harvest management technologies will give a further boost to Indian agriculture. While acceleration in agriculture growth to 4-4.5% is imperative, even with such growth rate; share of agriculture in total GDP is likely to reduce further. Therefore, there is a need to absorb excess agricultural labour in other sectors, notably industry. Rapid growth of agro processing industry close to the agricultural production centers can bring about this shift without moving people from rural to urban areas. Also, public investment in agriculture needs to be augmented, especially in rural infrastructure, irrigation, and agricultural research and development. Better access to institutional credit for more farmers, is also high on priority list. The New trade policy given pocus to agriculture and all the hurdles in Indian agriculture will be crossed gradually.

Impact of Economic Reforms Process on Indian Agricultural Sector

Agricultural sector is the mainstay of the rural Indian economy around which socio-economic privileges and deprivations revolve, and any change in its structure is likely to have a corresponding impact on the existing patterns of social equality. No strategy of economic reform can succeed without sustained and broad based agricultural development, which is critical for:

- Raising living standards,
- Alleviating poverty,

- ❖ Assuring food security
- ❖ Generating buoyant market for expansion of industry and services, and
- ❖ Making substantial contribution to the national economic growth.

Studies also show that the economic liberalization and reforms process have impacted on agricultural and rural sectors very much.

According to Bhalla (1997), of the three sectors of economy in India, the tertiary sector has diversified the fastest, the secondary sector the second fastest, while the primary sector, taken as whole, has scarcely diversified at all. Since agriculture continues to be a tradable sector, this economic liberalization and reform policy has far reaching effects on (i) agricultural exports and imports, (ii) investment in new technologies and on rural infrastructure, (iii) patterns of agricultural growth, (iv) agriculture income and employment, (v) agricultural prices and (vi) food security.

Reduction in Commercial Bank credit to agriculture, in lieu of this reform process and recommendations of Khusrao Committee and Narasingham Committee, might lead to a fall in farm investment and impaired agricultural growth. Infrastructure development required public expenditure which is getting affected due to the new policies of fiscal compression. Liberalization of agriculture and open market operations will enhance competition in "resource use" and "marketing of agricultural production", which will force the small and marginal farmers (who constitute 76.3% of total farmers) to resort to "distress sale" and seek for off-farm employment for supplementing income.

Scope and Objective of the Study

In this paper we have tried our best to highlight the role played by the agricultural sector in India in globalization era. For the purpose of our study, we have taken only secondary data collected from different journals. The period of the study is limited to two years, i.e. from 2002-03 and 2003-04. So, all limitations of the secondary data are found in this study.

Indian Agricultural Sector

The Indian Agriculture sector provides employment to about 65% of the labour force, accounts for 27% of GDP, contributes 21% of total exports, and raw materials to several industries. The livestock sector contributes an estimated 8.4% to the country GDP and 35.85% of the agricultural output. India is the seventh largest producer of fish in the world and ranks second in the production of inland fish. Fish production has increased from 0.75 million tons in 1950-51 to 5.14 million tons in 1996-97, a cumulative growth rate of 4.2% per annum, which has been the fastest of any item in the food sector, except potatoes, eggs and poultry meat. The future growth in agriculture must come from namely:

- New technologies which are not only "cost effective" but also "in conformity" with natural climatic regime of the country;
- Technologies relevant to rain-fed areas specifically;
- Continued genetic improvements for better seeds and yields;
- Data improvements for better research, better results, and sustainable planning;
- Bridging the gap between knowledge and practice; and
- Judicious land-use resource surveys, efficient management practices and sustainable use of natural resources.

IX Plan Strategy on Agricultural Development

The agricultural development strategy for the Ninth Five-year Plan is essentially based on the policy on food security announced by the government, to double the food production and make India hunger free in ten years. The strategy to ensure food security is as follows:

- Doubling food production,
- Increase in employment and incomes,
- Supplementary/sustained employment and creation of rural infrastructure through Poverty Alleviation Programmes (PAP),

- Distribution of food-grains to the people below poverty line (BPL).

The Ninth Plan Target is to achieve a growth rate of about 4.5% per annum agricultural output and production of 234 MT of food-grains by 2001-02. The policy thrust and key elements of growth strategy, as proposed in the Ninth Five-year Plan Document (Volume II: p. 444), are as follows:

- Conservation of land, water, and biological resources
- Rural infrastructure development
- Development of rainfed agriculture
- Development of minor irrigation
- Timely and adequate availability of inputs
- Increasing flow of credit
- Enhancing public sector investment
- Enhanced support for research
- Effective transfer of technology
- Support for marketing infrastructure
- Export promotion.

The Ninth Five-year Plan Document (1997-2002 Volume II) reveals that development of the vast rain-fed areas of about 90 m.h. would require over Rs. 37,000 crores. Further, scientific treatment for soil and water conservation for 12 m.h. of arable and 3 m.h. of non-arable land would require about Rs. 7,500 crores. Development of rain-fed areas require a substantial public investment, which may not be possible due to the new policies of fiscal compression. In the coming millennium on the basis of current trends in the consumption pattern, the estimated total requirement of food-grains is likely to be around 245 million tons by 2006-07.

Agricultural Planning and Development

India is a vast country with a variety of landforms, climate, geology, physiography, and vegetation India is endowed with regional diversities for its uneven "economic and agricultural" development, on account of (i) agro-climatic environments (15 zones/127 regions), (ii) agro-ecological regions (20) and 60 subregions, (iii) agro-edephic regions, (iv) terrain mapping sub-units, (v) natural resources endowments (geology, geomorphology, soil, groundwater, surface water, and infrastructure), (vi) human resources (Population density), (vii) level of investments in rural infrastructure, and (viii) level of investment in technology and its adoption.

India has a Total Geographical Area (TGA) of 329 million hectares (m.h.) out of which, about 265 m.h. represent varying degrees of potential for biological production. Report reveals that more than 50% of TGA is threatened by various types of land degradation, such as soil erosion, gully and ravine formation, salinity, water logging, shifting cultivation, etc. Development of irrigation potential is considered as the key factor in the sustenance of "Green Revolution". Despite 58 years of development planning, rain-fed agriculture is the largest and the most important sector of crop production in India.

Soil resources are the most precious non-renewable vital resources for growing food, fibre, and fuel wood to meet the human needs. Management of Soil Resources is essential for both the continued agricultural productivity and protection of environment. By considering various factors like population growth rate, diminishing per capita of land and water resources, and increasing land degradation problems, it is estimated that India will be required to produce an additional 5-6 million tons of food-grains annually in 21st century. This will lead to tremendous pressure on soil resources along with competitive demand for it from industrialization and urbanization. However, the capacity of soil to produce is limited and its limits to production are set by its inherent characteristics, agro-ecological settings, and its use and management.

Forests are an important natural resources of India, having a moderating influence against floods and also protecting the soil against erosion. About 95% of the forests in India is owned by states and the total area under forests is about 22% of the total geographical area.

Development of livestock has been envisaged as an integral part of sound system of diversified agriculture. In animal production, the major aim is for raising ecologically adapted animals and efficient utilization of locally available feed resource. Dairy development is intimately linked with cattle population, breed improvement, cattle health and disease management, and fodder development, etc. Animal Husbandry in India is essentially a endeavour of millions of small holders (Resource-Poor-Farmers) who rear animals on "crop residues" and "property resources" without generally allowing them to complete with man for food-grains.

The small-holders produce milk, meat, wool, etc., for the community, with virtually no capital resource, training and at a cost that no modern technology in the world had ever produced. Food and fodder resources will be crucial to the future development of "livestock resources" in the country. There is very little scope for increasing the area under fodder production, keeping in view the priority for food-grains, pulses and oil seeds. Development of Fodder Resources is basically an activity based on a multi-disciplinary approach involving the areas of agriculture, animal husbandry, environment and forests, revenue, rural development, and wasteland development.

Water resources of India contain diverse group of flora and fauna. Agriculture is the greatest user of water accounting for about 80% of all consumption.

Animal husbandry and fisheries require abundant water. Development of Water Resources, since independence, has been undertaken for specific purpose like irrigation, flood control, hydro-power generation, drinking water supply, industrial and various miscellaneous uses. Minor irrigation projects have both surface and groundwater as their source, while major and medium projects mostly exploit surface water resources. The break up of the ultimate irrigation potential under the above three categories is:

- 58 m.h. by major and medium irrigation projects,
- 17 m.h. by minor surface water schemes, and
- 64 m.h. by minor groundwater schemes.

Fisheries resources of India are either inland or marine. The principal rivers and the tributaries, canals, ponds, lakes, reservoirs comprise inland fisheries. The river extend about 27,200 km, and other subsidiary water channel comprise about

112,000 km. Marine resources comprise of about 2 Million sq.km of FEZ for deep sea fishing and 7,250 km of coastline. With the diverse fish fauna, the development objectives are to judiciously and optimally utilize the resources for (NBFGR, 2000):

- enhancing production and productivity of fishermen, fish farmers and fishing industry;
- increasing fish production and thereby, raising nutritional standard of people;
- earning of foreign exchange from export of marine products;
- improving socio-economic conditions of traditional fishermen;
- generating employment for coastal and rural poor; and
- conservation of depleting species of fish.

Good infrastructure helps in raising productivity and lowering the unit cost in the production activities of the economy. "Agricultural Infrastructure": refers to "Rural Infrastructure" whereas "Industrial Infrastructure" refers to "Urban Infrastructure". Agricultural development requires (i) agricultural research and extension, (ii) rural financial institution, (iii) irrigation and drainage, (iv) agricultural inputs (fertilizers, seeds, credits), and (v) marketing and storage facilities.

Agriculture credit is a crucial input for increasing agricultural production and productivity. Institutional finance for agricultural credit is disbursed mainly by Commercial Banks, Regional Rural Banks, Land Development Banks, and Cooperative Banks. Share of commercial banks in total institutional credit to agriculture is about 48%, that of cooperative banks is about 46%, and regional rural banks account for 6% only. Short-term credit account for 2/3rds of the total institutional lending to the agriculture.

Drought has multiplier effect on agricultural production during the subsequent year also, due to (i) non-availability of quality seeds for sowing of crops, (ii) inadequate draught power for carrying out agricultural operations as a result of either distress sale of cattle or loss of life, (iii) reduced use of fertilizers as the investment capacity of the farmers decline, (iv) non-availability of raw materials in agro-based industries, (v) de-forestation to meet the energy needs in domestic sector as agricultural waste may not be available in required quantity.

The Central Ministry of Agriculture (MOA) is responsible for implementation and formulation of national policies and programmes to achieve agricultural growth through optimum utilization of the land resources, water, soil, plant, fisheries, and livestock resources. Government of India implements the following agricultural related Schemes (whether watershed based or agro-climatic region based) in the country, which deal agricultural resources information for Planning and Development:

- Agro-climatic Regional Planning (ACRP) Project.
- Agro-Ecological Mapping Project of the National Bureau of Soil Survey and Land Use Planning (NBSS and LUP)
- All India Soil and Land Use Survey (AISLUS)
- Early Warning System of Agricultural Situation in India.
- Forecasting of agricultural output using space, agro-meteorology and land based observations (FASAL) Project
- Land Records computerization Project
- National Agricultural Research Project (NARP)
- National Agricultural Technology Project (NATP) to strengthen research-extension-farmer (r-e-f) linkage.
- National Watershed Development Program for Rain-fed Areas (NWDPRA)
- Soil and Water Conservation Programs
- Drought Prone Area Development programme
- Desert Development Programme
- National Wastelands Development programme
- Integrated Mission on Sustainable Development (IMSD) Programme.

Analysis

The analysis of the data of agriculture sector in India are made under two heads:

All-India Area, Production and Yield of Foodgrains from 1950-51 to 2003-04

The all India area, production and yield of foodgrains from the year 1950-51 to 2003-04 are explained in Table 1.

TABLE 1
All-India Area, Production and Yield of Foodgrains from 1950-51 to 2003-04

Year	*Area (million ha)*	*Production (MT)*	*Yield (kg. / ha)*	*% coverage under Irrigation*
1950-51	97.32	50.82	522	18.1
1960-61	115.58	82.02	710	19.1
1970-71	124.32	108.42	872	24.1
1980-81	126.67	129.59	1023	29.7
1990-91	127.84	176.39	1380	35.1
2000-01	121.05	196.81	1626	43.4

Source: *Report of Agriculture in India, 2003-04.*

Three Largest Producing States of Important Crops during the year 2002-03

Three largest producing states of important crops during the year 2002-03 are explained in the Table 2.

Conclusion

The central issue in agricultural development is the necessity to increase productivity, employment and income for poor segments of the agricultural population of whom the small and marginal farmers constitute a sizeable portion. Information Technology Tools viz., data warehousing (Data Bases and Model Bases), Expert Systems and Knowledge Bases, Networking (Internet, Intranet and Extranet), Geographical Information System (GIS), Application of Remote Sensing Data, Decision Support Systems, and E-Commerce (b2b, b2c solutions), facilitate the Farmers to know the "agricultural situation" in Indian as well as abroad and accordingly undertake agricultural production.

TABLE 2

Three Largest Producing States of Important Crops during the Year 2002-03

Crops / group of Crops	*States*	*Production (MT)*	*% share of production to all India*	*Commulztive % share of production*
		FOOD-GRAINS		
Rice	West Bengal	14.39	19.81	19.81
	Punjab	8.88	12.22	32.03
	Uttar Pradesh	8.11	11.16	43.19
Wheat	Uttar Pradesh	23.61	36.27	36.27
	Punjab	14.18	21.78	58.05
	Haryana	9.19	14.12	72.17
Maize	Madhya Pradesh	1.5	14.56	14.56
	Andhra Pradesh	1.49	14.47	29.03
	Karnataka	1.38	13.4	42.43
Toatl Coarse Cereals	Maharashtra	5.94	23.48	23.48
	Rajasthan	2.14	8.46	31.94
	Karnataka	3.63	14.35	46.28
Total Pulses	Madhya Pradesh	2.21	19.48	19.48
	Uttar Pradesh	2.06	18.49	38.33
	Maharashtra	2.05	18.4	56.73
Total Foodgrains	Uttar Pradesh	36.3	20.84	20.84
	Punjab	23.49	13.49	34.32
	West Bengal	15.53	8.91	43.23

Source: *Report of Agriculture in India, 2002-03.*

This IT-led globalization will certainly benefit the medium and large farmers who can invest on IT, as has happened during "green revolution". Since the agricultural development strategy has been mostly "growth-oriented" and therefore had a "build-in bias" in favour of large farmers over small farmers. Farmers can invest on computers to get access to Internet, but it is not possible for them to invest on "agricultural informatics" with decision support system using geomatics technology.

Agriculture being a "state subject" and a primary sector which accounts for about 27% of GDP, 65% of labour force, and 21% of total exports, the Central Government implements agricultural resources development schemes under both central sector and centrally sponsored sector. These schemes generate voluminous information, both spatial and non-spatial, related to agricultural resources, using conventional, remote sensing and GPS technology.

REFERENCES

Agricultural Report in India, 2002-03 and 2003-04.

Dhar, P.K., *Indian Economy and its Growing Dimensions*, Kalyani Publishers, New Delhi, 2004.

Dhingara, I.C., *Indian Economy*, Sultan Chand and Sons., New Delhi, 2004.

Economic Survey, Government of India, New Delhi, 2003-04.

Economic Survey, Government of Orissa, Bhubaneswar, 2004-05.

Rudra, Dutta and Sundaram, K.P.M, *Indian Economy*, S. Chand and Sons, New Delhi, 2004.

CHAPTER 2

Globalisation may Prospect Indian Economy, Retrospect Agricultural Produce in India

R.L. Panigrahy*

The topic will better be visualised in three phases as Phase I is Introduction to Globalisation, Phase II will be the prospects of Indian Economy in industrial sector including IT and Phase III will be how globalisation retrospect India's Agricultural Sector.

PHASE I

Introduction

Globalisation is the process of integration of world's economies in conditions of free flows of trade and capital and move and of persons across borders facilitated by new technologies. Internationalisation production has been taking place over the last few decades through the MNCs which operate with tens of thousands of affiliations.

As tariff barriers are getting lowered, trade is expanding, transport and communication cost falls, technologically advanced enterprises move to different locations, globalisation is turning the whole world into a common village. The era of globalisation gained momentum in 1990s. It has opened up new opportunities for developed and developing countries which are unable to form their own market and become internationally competitive.

*Lecturer, S.S. Polytechnic, Aska, Ganjam.

Not withstanding the risks and challenges flowing from globalisation, no single country nor even a group of countries even if they act together, would be able to arrest the march of liberalisation and globalisation. World trade has been growing faster than world output. Developing countries including India shared one of the fastest growth in world trade in 2002 of 12.5 %. Private capital flows of all forms to developing countries were 299 billion dollar in 2002. Of this, Foreign Direct Investment (FDI) flows to developing countries totally 178 billion dollars which have average of 170 billion dollars since 1996.

Advances in electronics and telecommunications have integrated the world's capital and financial markets to a large extent. There are no doubt visions of major developing countries like China and India becoming super powers in the 2lst Century. The World Bank has classified India along with China and other four countries as "Big Five", whose role in the world economy is likely to increase dramatically over the next two decades. Even with an average GDP growth of 5.8%, India's share of world real GDP would move up to 2.1% from the 1% in 1992. India's share of world export is expected to improve from 0.8% in 2000 to 3.9% by 2020. India's place will be determined by its ability to tackle poverty and unemployment effectively raise the level of education and health for the over one billion population.

PHASE II

Impact of Globalisation

Increased trade, new technologies, FDI, expanding media and internet connections are fulfilling economic growth and human advance. Globalisation now-a-days is being driven by market expansion which is out placing governance of these markets and their repercussions for people. More progress has been made in norms, standards, policies and institutions for open global markets than for people and their rights.

The challenges of globalisation in the new century is not to stop the expansion of global markets but one of setting rules and institutions for stronger governance - local, regional and national to preserve the advantages of global markets and competitions. But, also to provide enough space for human community and environmental resources to ensure that globalisation works for people not just for profit.

India's Export

In 2000, the India's export was 12% of global export. Due to global economic slowdown of 2001, it was declined to 1.5% in 2001. But export in April 2002 registered a healthy growth 19%. The Commerce Minister had visualised in its medium term export strategy for 2002-07. The economy will be better served if the Government focuses its energies on removing domestic bottlenecks and helping exporters identifying new products and markets.

FDI in India

FDI inflows during 2001-02 was $2.06 billion, a huge 66% increase over $2.46 billion received the previous fiscal year. In 1997-98 FDI was $3.56% billion. The most valuable form of FDI is that which goes into creation to new plants and establishments and expansion of equity in existing companies or for creation of new capacities A Government committee has been constituted to bring India's FDI estimation procedures in the global factors. In the last week of June 2002 Government of India (GoI) has welcomed more and more FDI in printing media. This is to be welcomed but a revised estimate of new and higher FDI is to of little value by itself.

Import Restriction

On 2002 Union Budget of India restricted on imports by increasing import duties and on agricultural products, second hand four and two wheelers and liquor products. That implies large proportion of products on which Quantitative Restrictions (QR) are to be removed on April will be subject to import tariff at the same rate as before. As the removal of QRs has been ongoing process, one should be able to measure the impact of earlier removal of import restrictions. A data of 1999 to 2000 is given below to better visualise the period and after QR situation of imports.

From the first group of data, the consumer goods imported immediately after QR, so import increased at a sudden. But in the second data imported are the consumables not quite necessary, so a low growth of imported. Anyway the import reduces, the foreign reserves a country. The rate of import increases in case of low or no foreign currency ratio. And also, the domestic product market will be in depression. Hence, Government should restrict import and product export.

TABLE 1
Products with QRs

	Lifted in April 1999		*Lifted in April 2000*	
	1998-99 QR era	*1999-2000 Post-QR era*	*1999-2000 QR era*	*2000-01 Post-QR era*
Value of Imports (increase)	11,802.00	13,821.00	5,151.56	2675.00
% increase of imports	6.6%	6.9%	2.6%	2.3%
No. of products imported	554	662	487	474
Value of Agricultural imports (in crores)	8123.00	8119.00	510.00	405.00

IT Export

The Electronic and Computer Software Export Promotion Council (ESC) is planning to launch a proactive programme to give software exporters and IT professionals proficiency in Japanese to help them focus increasingly on the expanding market in Japan. According to ESC Executive Director, D.K. Saren, the new market is poised to grow from $123.9 billion in 2002 to $150 billion by 2004, opened a mega opportunity to Indian software firms and professionals. He has estimated that software exports from India to Japan would go up to $ 2 billion in 2005 from $ 450 million in 2002. Mr. Kiran Karnik, President of NASSCOM (National Association of Software and Service Companies) has stated, about 34% of growth in software export will be occurred in 2004-05 in spite of 32% in last year. In 2003-04, India got $13.3 billion form software and service exports, but in 2004-05 it will reach at $17.9 billion. Government of India got Rs 46,100 crores as tax from software exports, a growth of 26% than previous year. In IT-enabled Service (ITES), IT products and technical know how, India got Rs.34,800 crores or $720 crores at 18.3 % growth than previous year. It education service and business earned 59% growth which earned tax of Rs. 11,300 crores or $ 230 crores by India. In

2002-03, the software business inside the country was Rs.12125 crores and projected $5000 crores in 2008. In comparison to geographical area America placed in first position and India is in the second position with 71% in software exports in the globe. Software Technology Park of India (STPI), a Government of India enterprise has produced 80% of software of India shared as shown in Table.

TABLE 2
Share of Software Exports (2003-04)

State	*Cost earned in Rs. in crores*	*Growth than previous year*
Karnataka	12350	25%
TN	6305	26%
Maharashtra	5508	20%
AP	3638	30%
Haryana	2734	28%
UP	2518	28%
Delhi	2065	18%
WB	1200	99%
Orissa	260	22%
Kerala	165	21%
Others	331	26%

Source : NASCOM Report 2004.

In the year 2003-2004, the increase of external trade increased 67% and internal trade increase 7.8%. Export of IT and Computers has been reached at Rs.46000 crores. So, Government of India made a export promotion policy that up to the export of Rs.2 lakhs, there is no export tax on IT and Software.

In the meanwhile India and Japan Governments are interested to work with partnership on Software and IT. Japan Government has eased visa regulations for IT professionals from India. Indian IT experts are now eligible for a multiple-entry visa with a three years validity. The overall IT-ITES (IT Enabled Services) exports from India will touch $ 17.9 billion in 2004-2005, a growth of 35% according to NASCOM Strategic Review 2005. The geographic spread of India's software export was expectedly US centric, with roughly 60% of the business coming from that country followed by Europe with 25%. Software Exports to Japan are expected to grow substantially, according to Electronic and Computer Software Export Promotion Council (ECS).

IT major such as Infosys, Wipro, Satyam, Hughes, Polaris, NIIT (National Institute of Information Technology) and nearly others 300 companies have set up offices and subsidiaries in Japan. Japan could look at outsourcing e-business export technology and architecture, engineering services, IT enabled services, quality assurance and embedded software work to India according to Nalin Kohli, Chairman, ESC. According to KBK Inforgraphics in March 2003, hiring of IT professionals was increased to 36% which was 8% in June 2002 and the professional employees will be shown in Table 3.

TABLE 3
Professionals of India from 1990, projected up to 2008

	Year						
	1990-91	*1996-97*	*1999-2000*	*2000-01*	*2001-02*	*2002-03*	*2007-08*
Professionals	56000	160000	284000	4330114	522250	650000	11000000

Source: NASCOMM, KRK Infosys.

PHASE III

The foreign crisis and shortage of supply necessitated the introduction of economic reforms in the middle of 1991. The three major components of reforms are Liberalisation, Privatisation, Globalisation (LPG). Globalisation ensure to enlarge competitive environment with backward and forward linkages which flow foreign capital and multifaceted technology to the domestic economy and access to foreign market. Globalisation also ensures transparency, flow of information and access to data and other information to one and all. It is good for few and worse others in case of developing countries like India. Because, industrial prosperity in India is very less and industrial degradations and retrospections are high.

India is an agro-based economy. Hence, Gandhiji rightly pointed out that India lives in village and agriculture. Hence, any type of agricultural development and related to will prospect Indian Economy. Any type of improvements of India in the global sphere like IT developments. Indian Human Resource is very qualitative, Indian people are very much efficient in foreign capitalistic organisations in the top positions, but in India they are useless. So, India needs the developments of infrastructure and rectifications of policies by government. After globalisation Government of India encouraged the farmers more to produce export quality products. But if we will survey elaborately about the figures of depressions of Indian farmers who are cultivating export-oriented produces like cotton, tea and coffee, flower and other sophisticated food items. But the miserable conditions like market failure and indebtedness to these farmers became a head ache of rural economy, it has many exemplary situations in India which are global debate. Hence, activists and developmentalists stated that the policy of an economy should look at the interest of the majority of the people. But, globalisation has made many loses of agriculture which is the harbinger of Indian economy. The tea and coffee export reduced in post-globalisation period than before. Other substitute market prevailed in the recent age. So, the slump in tea prices devastates the small farmers but MNC plantations can absorb several losses without going under. The plights of Andhra Pradesh weavers, driven to suicide, has been critically analysed by The *Frontline Magazine* (April 27, 2001). Textile policies since 1985 have sought to liberalise, modernise and privatise the industry. In fact these have succeeded in marginalising over 40 lakh handloom weavers who used to produce over 400 crores metre of cloth every year. Such a miserable condition in case of handloom and textile industry seen in all over the country. The farmers of AP are not traditional but advanced in multi-cropping pattern, sericulture, floriculture, horticulture with HYV seeds, fertilisers, pesticides for producing export quality of food grains and allied goods from agricultural wastes. The production needs more investment and they have to borrow which are mainly from money lenders or landlords at a gigantic rate of interest. But, many times they do not get the output profit oriented due to market failure, tax barriers, problems raised by agents, etc. which make them retrospect.

The scenario of Kerala is grim. Cooperatives came forward to upgrade farmers' livelihood. Organic farming is catching on. Until the green revolution dismissed farmers as illiteracy, ignorant and backward. The agriculture farming in Kerala is mainly coconut where the farmers are distressed to recoup their losses. Now the market driven policy has been made for coconut products.

In Orissa all the three spinning mills at Aska, Baripada and Sonepur (ABS) have been closed down showing heavy loss managed by the industrial department of Orissa. The hilly regions are of agricultural where there is no resource of water supply or groundwater, can be utilised in cotton cultivation. But, now Orissa is barren with cotton cultivation. Still few areas of Ganjam district have cotton cultivation, who sold to Andhra merchants of Kakinada and Vijayawada district at a cheap rate in which rate they do not get in their local market. In agricultural sector, the majority of production is paddy which has no market of sale at a right price but have highest retail price in the market. The crop failure due to natural calamities in each year and gambling of monsoon, single cropping in rainy season for four months, etc. most of the rural people migrating to urban and industrial areas of the state, neighbouring state and the states like Maharashtra, Gujarat, Tamil Nadu, etc. The irrigation facilities are only in pen and paper, but the farmers are paying water tax. The industries are closing down day by day. The farmers previously engaged in industrial sectors side by side which made their economy sound. Only the agro-based industry in large scale is sugar industry out of which two are closed since a decade and other five are open which produced 43 thousand tonnes of sugar where as state needs six lakh tonnes every year and the industries are in losses which are alarming to the farmers of the state. In all-India figures of sugar industries, most of the sugar industries are in miserable condition fighting for survival. But, the MNCs and big industrial firms interested to sell sugar in the market at a cheaper rate, which the worst attempt as a result of globalisation.

Before independence, the than Indian Government never think of the lacunas of globalisation. Now, both the Central and State Governments are trying to maximise the profit of their departments. They have minimise their expenses, squeezed the strength of staff under VRS and CRS scheme and not filling the posts lying vacant by retirement. After such cares taken, the only alternative is to privatise. Due to privatisation of many departments, the company will think of maximum profit but utility service. Before globalisation, many departments are meant for public utility. But nowadays almost all the departments are privatised and trying to maximise profit not the greater interest of the public. Where India is still poverty ridden (40%) and majority (80%) of Indian do not cope with sudden rise in price, mass illiteracy (300 million people), no safe drinking facility (1,75,000 villages), 80% student drop out at 8th standard and 50% of others failing in SSC standard, 90% of students deprived of any opportunity to earn their livelihood, prevalent socio-economic deprivation, prevailing starvation death, mass unemployment, capitalistic pattern of society instead of socialistic, etc. are the alarming aspects of Indian economy which are the pressure to the agriculture and the problems can be diminished by agricultural developments which the globalisation never fulfil. These problems can not be realised by the rich people because globalisation made them faster in earning.

The latest Human Development Report pointed out that India has one of the world's lowest expenditure on human priority concerns such as primary health and education, safe drinking water and family welfare. It seems that there is inconsistency on the performance of economic reforms. The foreign investors are trying to establish their business in India due to easy and cheap availability of raw materials and quality human resource. As they are easy and cheap in establishment, they are showing their less interest with the demand to minimise tax, complex legal procedures to open and close their establishments. In this case, they will rush to our economy, demolish our establishments at lowering rate of their quality and sophisticate products, try to maximise their profit by their cleaver policy of marketing and if fails they will fled, away easily. In this situation, the closed down domestic establishments never raise their heads again.

Trade liberalisation has been touted as the magic key which would open the doors to economic prosperity. We told that, it is the only panacea for all the ills to which our economy is prey, from low productivity to unemployment to obsolete technology. Exports are now painfully inching their way up but imports have overtaken them with remarkable speed. While the wealthier nations of the world are busy forging their own trading blocks to ensure a better bargaining power, developing countries, including India are twiddling their thumbs, waiting for men to fall from heaven. Yet there are a few shrill small voices crying out in the wilderness against the pitfalls of reckless liberalisation.

No body bothers about them who feed the world but live in perennial fear of the weather, crop disease, markets and a host of other unpredictable variables. The economic efficiency here seems to imply that the poor are inefficient rather than spelling out the odds against them, that the factors that keep them poor. The progress of the poor can not take place without displacing the rich who exploit them. Farmers are in much the same condition. Prices of agricultural produce are crashing, power tariffs have been reduced or removed. Yet retail price remains high, allowing only the middlemen to make a profit. (Journalist P. Sainath stated in *The Hindu*, Feb 25, 2001). It is not natural calamity but the cruel, heartless policy of a decline in investment in the agricultural sector which has brought about the lowest rate of growth. Fads are imposed upon the farmers without giving the matters of serious thoughts.

Conclusion

When any body arises against the governmental policies on globalisation and economic prospects, industrial developments, why Indian industries raise big questions on facing challenges of operating in a deregulated and competitive economic environment. If Indian industries are too comfortable, why they has been lobbying very hard with the government to reject or put difficulties in the way of foreign multinational or collaborations or investments coming India in a big way? A large section of industry is in dilemma.

Given the irreversibility reforms and continuing reforms in tariff interest and exchange rates policies, the Centre-State partnership should grow to facilitate export and to promote country's basic strength. Labour and capital productivity should be raised. Assimilation and innovation with regard to technology would need promotion of research and development in a big way. Along with economic reforms, there is a great need for administrative reforms. There is no gain saying the fact that economic reforms can be sustained only through administrative reforms. This thing very remote in India because of red tapism and the bureaucratic hassles. Therefore, the process of reforms need to internalised. Special care should be taken to make progress of the poor who can not speak against the person who exploit them, they are deprived of livelihood from agriculture, sincere, hardworking, thinking of their fortune after harvesting, illiterate, ignorant. Gandhian focus on the India's villages had been widely accepted, in theory at least. Gandhian economics is best suitable for the Indian economic i.e. reforms on domestic products with self-reliance than at hardship depend on others. But in globalisation our economy depend on external investment/products not think of domestic. Because, we think of our environment, culture and tradition, on which we can manage our economy at best but without corruption at bureaucrats.

REFERENCES

Ahmed, S.N., Globalisation and Decentralised Development, *Kurukshetra*, July 2003, p. 4.

Bhattacharya, B., Globalisation competitiveness of Indian agriculture, *Yojana*, 2002, p. 21.

Chandhake, N., 'The sweep of Globalisation', *The Hindu*, 1.7.2002, p. 10.

Course in Japanese for IT professionals, *The Hindu*, 22.9.2002, p. 9.

'Diffusion of IT, is it so?', *Chanakya Civil Service Today*, June 2004.

Economic globalisation has become a war against nature and the poor, *Chanakya Civil Service Today*, Jan 2005, p. 74.

Economic reforms - How far and what further?, *Chanakya Civil Service Today*, Oct 2005, p. 34

'Ensure your success with the hottest IT programme as IT jobs make a comeback', *CSR*, Sept. 2003, p. 26.

Globalisation to be fit few, *CSR*, Aug 2004, p. 16.

Improvement of 34% in IT export in 2004-05, *Orissa Bhaskar*, Mar 2005, p. 10.

Kannan, S., 'Software potential in yen countries', *The Hindu.* 19.3.2005, p. 16.

Koshi, N., *Globalisation the Imperial thought of Modernity*, Vikash Adhyan Kendra, Mumbai.

Panigrahy, R.L., 'India in the field of software development', *Dainik Asha*, 08.2.2004, p. 4

Panigrahy. R.L. "Sugar Industries of Orissa A Study". In: *Industiral Prospects of Orissa*", J. Pradhan (Ed.), Bhubanesware (2005).

Raul, R.K., BPO in Indian landscape, *Yojana*, May 2004, Vol. 48, p. 32.

Rout, R., Globalisation and India's future, *Orissa Bhaskar*, 29.1 .2005, p. 5.

Schenke, T., 'Globalisation of world economy', *German News*, March 2002, p. 7.

Sustainable livelihood, *Chanakya Civil Service Today*, Jan 2005, p. 46.

Trade and Investment, *The Hindu*, 24.6.2002, p. 10.

CHAPTER 3

Globalisation and Sustainable Agriculture in India

Dr. Dasarathi Bhuyan

Globalization, no doubt, relates to the globe or world. In reality, it is a new outlook and a new attitude of looking at things in terms of the world instead of a single country. Though globalisation, liberalization and privatization are three different words and have different meanings, yet their basic purpose is common and that is to remove the unwanted restriction on economic forces and from economic view point to convert the world into a 'Global Village'. In general we can say that globalisation refers to a process by which the planet earth is considered to be one single unit or a "global village", where social and economic interactions among the people are based on interdependence. The word is considered to be a global society with global issues and problems which are to be tackled with global efforts and cooperation. Various scientific and technological developments have brought the world quite close. Globalisation is also manifest in the rapid flow of information, capital and goods. Under the impact of globalisation cultures and societies that were hitherto district have come face to face with each other. Globalisation originates from developed countries and technologies, capital, product and services come from them to developing countries. It is for developing countries to accept these things adopt themselves to them and to be influenced by them. As a result, the values and norms of developed countries are gradually rooted in developing countries. This leads to the growth of a monoculture. The culture of the developed countries being imposed on developing countries.

The contemporary, world is faced with dangerous problems, such as, environmental degradation, increased use of pesticides has emerged as a potential source of danger to sustainability of environment that endangers the existence of all forms of life on this planet. Perils and pitfalls of pesticides have been well evidenced due to their residual toxicity in our food chain. From a number of trials conducted across the country, toxic residues of pesticides have been revealed in the food stuff not only of plant but also of animal origin like milk and milk products, fish, meat, egg etc. at concentrations much higher than the permissible level of human body. Some scholars have argued that globalization created this problem in the developing countries like India. This is mainly due to the economic globalisation, globalisation of skill, knowledge and technology. Knowledge about production methods, techniques, and economic policies are available at a very low cost after globalisation. It ultimately become a highly valuable resource for the developing countries. But these countries are choking on its own agricultural success.

Therefore, the apparent contradiction of our necessity for nutritional security on one hand and environmental sustainability on the other makes it inevitable to resort to the organic or eco-farming system as it appears to be a possible option to meet both these objectives. The sustainable agriculture implies a farming system that primarily aims at cultivating land and raising crops under ecologically favourable condition. It emphasizes restricting the use of chemical inputs whether it is inorganic fertilizers or pesticides and instead, relies more on an integrated approach of crop management practices making use of cultural, biological and natural inputs.

What is Sustainable Agriculture?

Sustainable agriculture has been conceived differently by different persons and authorities. According to Food and Agriculture Organization Report of 1989 sustainable agriculture mainly involves the successful management of resources for agriculture to satisfy changing human needs while maintaining or enhancing the quality of the environment and conserving natural resources. In a nutshell it refers to an agriculture that aims at meeting the current food demand of the present generation without endangering the resource base for the future generation. In other words sustainable agriculture envisages successful management of resources for agriculture production to satisfy changing human needs, while maintaining or enhancing the quality of the environment and conserving natural resources.

The American Society of Agronomy defines, sustainable agriculture as one that over long term (1) enhances the environmental quality and the resources base on which the agriculture depends, (2) provides for human fibre and food needs, (3) is economically viable and (4) enhances the quality of life for farmers and the society as well.

Thus, it emphasizes on increasing trend in productivity without tempering with the sustainability and integrity of the biosphere. It relies on diverse inputs to maintain the ecological balance and enhances economic viability of the farming. The Technical Advisory Committee of the Consultative Group of International Agricultural Research (CGIAR) has also emphasized that sustainability in agriculture, agricultural researches on global basis must be directed towards, maintaining agricultural production at levels necessary to meet the increasing needs and aspiration of the expanding population without degrading the environment.

Sustainable agriculture includes economically viable system that reduces use of the off farm inputs such as chemical fertilizers and pesticides and relies more or on-farm resources. The goal of sustainable agriculture is to feed the expanding population while farming profitably in an ecologically sound, regenerative way. In order to be sustainable, agriculture needs to be (i) technologically feasible (ii) economically viable, (iii) socially acceptable and (iv) environmentally sound. Sustainability in agriculture can be achieved broadly through - efficient management of natural resources like land water and forest. And integrated approaches to crop management.

Organic Manures in Organic Farming

Organic manure in broad sense includes composts from rural and urban wastes, crop residues, agro-industrial bio waste and green manures; apart from the commonly used FYM. Organic manure improves soil physical condition including soil porosity and water-holding capacity and microbial environment.

Organic manure is of paramount importance not only in augmenting the crop production but also for making the agriculture sustainable as an eco-friendly means of soil health management.

Vermicoposting

Vermicomposting is an effective means of composting the decomposable organic wastes using earth worms naturally present in the soil. Vermicomposing is a mixture of worm casts enriched with macro and micro-nutrients. Vermicompost improves the physical and biological condition of soil, improves soil fertility and pulverizes it through their churning and turning action in addition to contributing plant nutrients, improves aeration and water-holding capacity.

Exploitation to Biological Nitrogen Fixation

Atmosphere containing as much as 78% nitrogen can be a potential source of this essential nutrient to soil utilizing the natures' own mechanism of biological nitrogen fixation. It not only offers an economic and ecofriendly source of nutrient supply, moreover evidences show that nitrogen so derived is less prone to loss than fertilizer nitrogen and as a long term effect it builds up a reserve of readily – mineralizable organic nitrogen. Thus it plays an important role in sustainable agriculture.

Minimum Plough and Tillage

The conventional tillage operation has disastrous effect on soil erosion causing greater loss of nutrients. This has led to necessity of stubble mulching. In the stubble mulching the soil is protected by the crop residues left on the soil surface during fallow periods. This has now become well established that instead of frequent tillage operations minimum tillage is more useful not only because it offers cost effectiveness but also contributes on conservation of soil and moisture.

Selection of Crops for Crop Rotation

Crop rotation has an important role to play in organic farming. Judicious selection of crops in crop rotation helps in efficient utilization of plant nutrient from different depths of soil. The practice of inter-cropping also helps in minimizing the risk of crop failure due to uncertain rainfall and infestation of pest and diseases.

Integrated Pest Management

The multifarious harmful consequences of indiscriminate use of pesticides have cropped up as a serious threat to the ecosystem. In view of the fact that increased use of pesticides has been drastically endangering the environmental sustainability, integrated approach to pest management needs adequate importance to make the agriculture ecofriendly.

Water Resources and its Management

About 72 per cent of cultivated area of the country depends entirely upon rainfall. Hence, efficient rainwater management plays a vital role in ensuring stability and sustainability in agricultural production. As it is the natural resource that cannot be created as per desire, efficient management of the water resources is the need of the day.

Land Resources and its Management

The constant increase in the demand of the bourgeoning population for food, fodder, fuel and shelter puts a tremendous pressure on our land resources always resulting in a continuous decline of the cultivable land area at a very fast rate. Efficient land resource management needs to be given adequate attention to increase the productive capacity of land and to prevent it from deterioration. Suitable location specific soil conservation and land reclamation measures based on soil survey on watershed basis needs due priority.

Management of Forest Resources for Sustainable Agriculture

Shifting cultivation, a primitive system of agriculture is prevalent in tribal areas of India. The soils are poor, infertile with low water-holding capacity situated on sloppy land. This system involves the cultivation of crops on a patch of cleaned forest area vicinity of their settlement. Trees and bushes are cut during November to January, allowed to dry and burnt on fire. Before the onset of monsoon seeds are sown or dibbled. In on patch, this process continues for 2/3 years till the fertility runs down. Then, the patch is abandoned and new site is selected for the purpose. The abandoned patch regenerates, become fertile but the original forest flora, fauna and ecology are never restored. It is an easy method of deforestation. The wild animals loss their shelter. Springs below the hills dry up, causes heavy flood in the rivers below. Increases water scarcity and moisture scarcity for plants, animals and human beings.

The problem of soil erosion due to shifting cultivation is very serious in states like, Orissa, Andhra Pradesh, Jharkhand, Chhattishargh. The system of shifting cultivation cannot be avoided completely but should be discouraged. The alternate substitute for the system should be advocated among the tribals. It may not be possible to stop shifting cultivation overnight but the tribals can be persuaded and educated to adopt modern technologies for their resources on a sustained basis.

Effects of Globalisation for Sustainable Agriculture

Due to the globalisation of awareness many general problems are solved. Similarly, this awareness would be a positive impact for the achievement of sustainable agriculture. A new consciousness can be generated in the whole world for sustainable agriculture. There should be international agreements on environmental standards, and on use of pesticides. All states should abide by them. Environmental standards should be imposed by the international community which hinders the path of sustainable agriculture.

Globalisation is capitalism in its globalized form. The end of the cold war marked the victory of capitalism over communism. Capitalist democracy has decisively won, and communism has accepted defeat. As capitalism advocates non-interference of the state in the individual life and non control of means of production by the state, free competition among the farmers will lead to more production. In this process over exploitation of natural and renewable resources and indiscriminate and irrational use of synthetic inputs like inorganic fertilizers and pesticides in view of producing more and more form unit piece of land are being increasingly realized to seriously impair the ecological balance and putting the environment in jeopardy. In the name of economic development globalisation would certainly destroy environment and sustainable agriculture. Are these signs of the time? But eventually, the world community may find that it will be cheaper to use pesticides, genetic potential of plants, scientific inputs and technology now rather that to pay for its nasty ill effects later.

REFERENCES

Baral, J.K., S.C. Hazari and A.P. Padhy, *Globalization in Foundations of Politics and Government*, Vidyapuri, 2004, p. 61.

Dhrubanarayan, U.U., *Wasteland Management*, ICAR Publication.

Gopalan, Radha, Sustainable Food Production and Consumption, *Economic and Political Weekly*, April 14, 2001, p. 1207.

Lenka, D., *Agriculture in Orissa*, Kalyani Publishers, New Delhi, 2001.

Mishra, B.B. and K.C. Nayak, Organic Farming for Sustainable Agriculture, *Orissa Review*, Oct., 2004, p. 42.

Mishra, B.B., Sustainable Agriculture for Food Security, *Orissa Review*, May, 2005, p. 34.

Sahu, S.K., R.K. Nayak and D. Sarangi, Sustainable Soil and Land Management under Shifting Cultivation in Orissa, *Orissa Review*, January 2005, p. 49.

CHAPTERT 4

Agricultural Credit Through RRBs: A Study of Regional Rural Bank in India

Bhagaban Pandhy*

India's economy has been in the past, currently and will remain in the foreseeable future predominantly rural in character. This is evident from the very high proportion of India's population living in rural areas. Even today majority (around 70%) of the total population live in villages and about two-thirds of its work force is engaged in agriculture and allied activities in rural areas. However, the British didn't give due importance to agricultural development and completely mined the rural economy of India by initiating the process of industrialization and giving utmost importance to it. After independence the situation was aggravated as Nehruvian concept of industrialization was followed and Gandhiji's vision of rural India was neglected. Keeping in view the industrialization, in the first two Five-year Plans importance was given to rapid industrialization. This led to lope sided development which compelled the planners to feel that Indian economy largely, depends on agriculture. Without its development Indian economy cannot grow and credit is highly essential to lead the agrarian/rural economy of India. Besides globalisation has brought a dimensional change which needs a strategic discussion. Because supply of credit, plays very important role for the survival and growth of argarian economy of India.

---

*Lecturer in Economics, A.N. College, Dharakote, Ganjam, Orissa.

Credit is an important input in the process of development and an accelerator of development that is why ensuring the provision of timely and adequate credit to the large segment of the rural population has been one of the major policy initiatives of the present day planning in India. Credit needs of the rural people arise out of productivity as well as unproductive purposes. Such requirements emerge from the various rural economic activities like agricultural operations both under traditional and modern methods, processing and marketing of agricultural products and undertaking allied activities. And today, rural credit in India is being provided by three agencies viz, Co-operatives, Commercial Banks and Regional Rural Banks (RRBs) where RRBs play very significant role.

Methodology

This study has been done by taking into account the secondary data in relation to the growth of RRBs and their Branch network, coverage, volume of business, advances and deposits, profit and loss etc. the data have been analysed by employing simple statistical tools such as averages, percentages etc.

Reasons for an alternative credit source in India:

(a) Non-availability of timely and sufficient agricultural credit by the institutional agencies.

(b) En-route of savings from rural to urban centers by the commercial bank.

(c) Lack of sufficient funds for agricultural development from by the commercial banks.

(d) To save the farmers from the clutches of moneylenders providing conditional loans at exorbitant rate of interest.

(e) Profit motive and high cost structure of the commercial banks.

(f) Unjust distribution of funds by the commercial banks.

That is why a number of committees, study groups and study teams were appointed by the Government of India and the Reserve Bank of India to study and recommend suitable measures to meet the rural credit requirements among these Gadgil Committee, Talwar Committee, Hazari Committee,

Tandon Committee, Kamath Committee, Desai Committee are worth-mentioning. And most of them have expressed, time and again, the need of rural institutional credit system for rural development. However in 1972, Sariya Committee recommended for the creation of new credit agency namely "Rural Banks" to meet exclusively the credit needs of the rural poor. With a view to find the feasibility of such Rural Banking and their functioning the Government of India appointed a committee under the Chairmenship of M. Narasimham. This committee in its report strongly recommended for setting up of such Rural Banks which should be rural and regional based but fully owned by the government. These banks are to work as supplementary credit agencies to the existing ones. As per the recommendation of the committee the Government of India promulgated the Regional Rural Banks ordinance on September 26, 1975, which later became the Regional Rural Banks Act, 1976 and was made effective from the date of the ordinance.

RRBs in India

As per the ordinance of 26th September 1975 initially a total of five RRBs have been set up at Moradabad and Gorakhpur in Uttar Pradesh, Bhiwani in Haryana, Jaipur in Rajasthan and Maida in West Bengal. These banks were sponsored by the Syndicate Bank, State Bank of India, Punjab National Bank, United Commercial Bank and United Bank of India. And in due course the number has came upto 196 as on March 2005 with more than 14,500 branches. Each RRB is sponsored by a nationalized bank known as a sponsoring Bank, which provides all sorts of help to these RRBs. The area of operation of these banks is specified. The jurisdiction of each RRB is confined to specified districts in a State. The number of districts varies between one to five and each branch office covers one to three blocks covering at least five to ten farmers service societies.

Objectives of the RRBs

The main objectives of RRBs are as follows:

(a) to mobilize deposits from its regional rural area and deploy them in the same region;

(b) to have a key role in lending productive credits to the weaker sections of the society comprising small farmers, marginal farmers, land labourers, persons engaged in trade, commerce and in any other productive activity;

(c) to provide consumption less loans to the needy rural poor up to a certain limit. The agricultural credit given by RRB, may be grouped under three categories.

Profile of RRBs (As on March 2005)

Total No. of RRBs	196
No. of RRBs that made profit	167
Deposits	Rs. 62143 crore
Borrowings	Rs. 5524 crore
Investments	Rs. 23200 crore
Total Assets	Rs. 77866 crore
Outstanding Advances	Rs. 32871 crore
Operating Profit	Rs. 1009 crore
Net profit	Rs. 750 crore

Direct Agricultural Advances

The financial assistance granted to farmers and landless agricultural labourers are treated as direct agricultural advances and such loans can be given to eligible persons as investment credit for the creation of fixed assets or production credit for meeting seasonal requirements. Thus RRBs are expected to take up the following schemes:

- ❖ Scheme for crop production credit.
- ❖ Scheme for loans against pledge of cold storage receipt.
- ❖ Scheme for loans for irrigation.
- ❖ Scheme for land development.
- ❖ Scheme for equipments and implements.
- ❖ Scheme for cart and draft animals.

Allied Activities

Apart from the agricultural credit these are expected to provide loans for allied agricultural activities viz., dairying, poultry, piggery, sheep and goat rearing etc.

Indirect Agricultural Advances

The RRBs may also finance the Primary Agricultural Co-operative Credit Societies (PACS) and Farmers Service Societies (FSS), which have a preponderance of small and marginal farmers, landless agricultural labourers etc.

In this backdrop attempts have been made in this paper to explore the following objectives:

The paper has the following objectives:

1. To examine financial support to agriculture and allied sectors by rural Banks.
2. To know about the constraints faced by RRBs in financing the agricultural needs.
3. To suggest some further policy measures.

Analysis

This section intends to present the entire study in three parts:

PART I - financial support to agriculture and allied sectors

PART II - constraints faced by RRBs in financing the agricultural needs

PART III - Suggestions and Concluding Remarks

Table 1 shows that the number of RRBs has remained constant since 1990. Only the number of branches which has gone up marginally, is not sufficient to support the vast agrarian economy of India. Deposit position is also not very significant. Table 2 shows that co-operative institution which are struggling for survival in India have been playing very important role in providing institutional credit to agriculture in India. So also Commercial Banks but the role of RRBs is not significant which is probably due to lack of funds. Table 3 shows the finance provided to agriculture by different financial institutions. The role of RRBs are appreciable in comparison to that of other financial institution. Table 4 shows the improvement in the recovery by the RRBs starting from 1993 to 2003. No doubt the recovery position is sound, but not sufficient so far as the position of others financial institutions are concerned.

TABLE 1
Expansion of RRBs during 1975-2003

Period Ending	*Banks*	*Branches*	*Loans (in Rs. Crore)*	*Deposits*	*Period Ending*	*Banks*	*Branches (in Rs. Crore)*	*Loans (in Rs. Crore)*	*Deposits*
Dec 1975	6	17	0.10	0.20	Mar 1988	196	14,508	8,486.62	19,325.65
Dec 1980	85	3,279	243.38	199.83	Mar 1999	196	14,508	9,367.21	23,597.61
Dec 1985	188	12,606	1,407.67	1286.82	Mar 2000	196	14,311	13,814.89	32,204.34
Mar 1990	196	14,443	3,554.04	4,150.52	Mar 2001	196	14,311	15,816.30	38,271.87
Mar 1995	196	14,509	6,290.97	11,150.01	Mar 2002	196	14,390	18,629.22	44,539.15
Mar 1997	196	14,508	7,852.66	15,423.42	Mar 2003	196	14,433	22,157.85	50,098.34

Source: Annual Report NABARD (2002-03).

TABLE 2
Flow of Institutional Credit to Agriculture

(Rs. in crores)

Institutions	*1997-98*	*2000-01*	*2002-03*
Cooperative Bank Share (%)	14085 (44)	20,801 (39)	24,296 (34)
Regional Rural Bank Share (%)	2040 (6)	4,219 (8)	5647 (8)
Commercial Bank Share (%)	15831 (50)	27,807 (53)	41,047 (58)

Source: NABARD.

TABLE 3
Micro-finance in India

Bank	*Comulative No. of disbursement per SHGs provided*			
	Bank loan SHG			
	Up to Mar 01-02	*Up to Mar. 04*	*During 2001-02*	*During 2003-04*
Public Sector Banks	118855	516697	0.020	0.057
Private Sector Banks	5391	21725	0.051	0.155
Regional Rural Banks	84775	405998	0.023	0.013
Cooperatives	12773	134671	0.019	0.036
Total	221794	1079091	0.021	0.051

Source: NABARD, 2004.

Problems Faced

RRBs face multifarious problems due to their peculiar nature and tasks involved. The main issues are as follows:

1. Poor recovery rate and the resultant, mounting losses.
2. Delay in decision-making on account of different agencies being involved in the management process.
3. Capital inadequacy.
4. Staff incompatibility as they find it difficult to cope with rural surroundings and business opportunities.

5. Restrictions in respect of deposit mobilization and scope for investment.

TABLE 4
Recovery performance of RRBs (as on 30th June, 2003)

(Rs. in crores)

Year	*Demand*	*Collection*	*Balance*	*Recovery %*
1993	2,726.65	1,123.49	1,603.16	41.20
1994	3,159.95	1,460.85	1,699.10	46.20
1995	3,669.07	1,870.36	1,798.71	51.00
1996	4,426.88	2,439.18	1,987.70	55.10
1997	5,503.29	3,143.08	2,360.21	57.10
1998	5,890.29	3,559.28	2,331.01	60.42
1999	7,034.23	4,519.30	2,514.93	64.24
2000	8,026.36	5,473.59	2,552.77	68.20
2001	9,617.93	6,789.53	2,828.40	70.59
2002	11,569.82	8,274.34	3,295.34	71.52
2003	13,246.95	9,738.80	3,508.15	73.53

Source: Annual Report NABARD (2002-03)

Suggestions

In the light of the foregoing analysis and the report of the study group on RRBs, the following suggestions are made, considering the need for strengthening intermittent supply of rural finance and effective operation of the RRBs in the days to come.

Increasing in Branch Network

Increasing of the number of branches in general and rural branches in particular sidelining the recommendations by Kelkar Committee not to open new branches.

Owners and Control

Instead of these banks working as extensions of sponsor banks, they must be made to work independently as a separate entity.

Target and Motive

Advances target should be fulfilled at any cost. Almost all the advances must go to the rural sector. There must be flexibility in the interest rates between different kinds of loans such as crop loans, village housing loans, cottage industries loans.

Coverage Area and Branch Expansion

India has about 5 lakh villages. As the scheduled banks have already diverted their attention to urban and metro cities resulting in closure of many rural branches, the only alternative is RRBs, barring the co-operatives. It is essential that all these villages, panchayats, blocks, taluks must come under these Banks. It is essential to multiply the number of branches to cover the entire rural area of our country.

Staffing and Training

The appointment and training of the staff working in these rural-0oriented institutions should be given much attention. Every branch should have on its staff, at least one person from the same locality so that it will be easy for the Bank Manager to learn about the local conditions and people. Generally, members from rural communities only should be appointed in these banks so that the employees do not appear to be out of place in the rural setting. Also they can serve better in the familiar surroundings.

As Market Leader

The RRBs should have the status of the market leader of rural finance. The number of other players must be restricted.

Conclusion

As institutional finance is inadequate in rural areas, moneylenders rule the roost and the rural population is ever in the shackles of debt and bondage. The commercial banks are closing down rural branches and concentrating on profitability rather than service to the rural poor. Organized lending through government institutions is to be reinforced to help the villages recover from debt and poverty and improve employment and productivity. In this context, the RRBs are specially suited to work solely in the rural areas to channelise funds and monitor regional development.

REFERENCES

Dhingra, I.C., *The Indian Economy: Resources Planning Development and Problems*, Sultan Chand & Sons, New Delhi.

General Studies, Indian Economy, *Pratiyogita Darpan*.

Kurukshetra, Vol. 51, No. 10, August 2003.

Pany, R.K., *Indian Economy*, Kitab Mahal, Cuttack.

RGB (Head Office Berhampur) Annual Reports: 15th, 17th, 18th, 22nd, 23rd & 24th.

CHAPTER 5

Genetically Modified Crops: Commercial, Ethical and Environmental Issues

Ashok Munjal,* Joginder Singh Thakur# and Ravi Kant Pandit+

ABSTRACT

Like most human planetary management issues today, such as global climate change, the GM foods issue is hugely complex. GM foods have great promise and great dangers. If all goes well, one thing is certain, we will have to feed about 12 billion people everyday in the next 50 years. Good and evil are moral choices humans are free to make. Twenty-first century choices face us in stem cell research, cloning and genetically-modified foods. Proponents and opponents present their cases and policy-makers are faced with protecting the public interest. It is within this context that we look at the case for genetically-modified, engineered organisms and foods. Only two traits, herbicide and insect resistance have been of significant commercial success. Crops with other traits have failed to achieve any commercial success, have been held back by companies, or never made it through the research and development pipeline. It would be a serious mistake to over understand the positive early experience with *B.t.* and herbicide-tolerant crops and conclude that the weak regulation

* Reader, Department of Bioscience and Biotechnology, Banasthali Vidyapith -304 022 (Rajasthan).

Sr. Lecturer, Department of Botany, Government College, Mandi (Himachal Pardesh).

+ Sr. Lecturer, Department of Zoology, BBN PG College, Chakmoh-176039 (Himachal Pardesh).

currently in place will suffice to control the risks of these and other new crops. Within the next few years literally thousands of genetically-modified food products or products with genetically-modified components could be put onto the world market. The incentive for pursuing this area of biotechnology comes in four main categories: better health, better products, better for the environment, and better business. There are a number of practical and ethical issues associated with this technology that is not readily understood by the public and often become emotive when publicized through the news media. It is important for people to be able to knowledgeably discuss the possible benefits and costs of this technology along with some of the ethical concerns as public debate heightens in the future.

Introduction

The terms genetically-modified (GM) or genetically engineered (GE) foods and genetically-modified organisms (GMOs) refer to crop plants and organisms created for human or animal consumption. GM is a special set of technologies that alter the genetic make-up of such living organisms as micro-organisms, plants, or animals manipulated at the molecular level to have traits that farmers or consumers desire using the latest molecular biology techniques, recombinant DNA technology etc. or we can say "To isolate single genes of a known function from one organism and transfer a copy of that gene to a new host to introduce a desirable characteristic". These foods often have been produced by the insertion of "foreign" genes taken from sources other than the organism's natural parents. So, GM plants contain those genes they would not have contained if researchers had used only traditional plant breeding methods. These techniques of modern genetics have made possible the direct manipulation of the genetic make-up of organisms. The resulting organism by combining genes from different organisms is said to be "genetically-modified," "genetically-engineered," or "transgenic". Much of the food consumed in the United States is genetically-modified.

This technology can be used to produce new varieties of plants or animals more quickly than conventional breeding methods and to introduce traits not possible through old traditional techniques. Recent innovations in biotechnology allow scientists to select specific genes from one organism and introduce them into another to confer a desired trait viz., such as those conferring insect resistance or desired nutrients is one of the most limiting steps in the process. The principal agricultural biotechnology products marketed to date have been genetically-modified crops engineered to tolerate herbicides and/or resist pests. Glyphosate and glufosinate-ammonium are the two main herbicides for which crops have been modified for herbicide tolerance (Betz *et al.*, 2000). Crops carrying herbicide-tolerant genes were developed so that farmers could spray their fields to eliminate weeds without damaging the crop. Likewise, pest-resistant crops have been engineered to contain a gene for a protein from the soil bacterium, *Bacillus thurigiensis.* This protein, referred to as *B.t.*, is produced by the plant, thereby making it resistant to insect pests like the European Corn Borer or Cotton Boll Worm. Other pest-resistant GM crops on the market today have been engineered to contain genes that confer resistance to specific plant viruses. So this technology enabled us not only to develop insect/disease resistant but also to develop such varieties that are suited to specific consumer, industrial etc. need are also popularly called "Designer Crops".

Scientists are able to extract DNA/gene from any organism and can then isolate a specific gene through the use of restriction endonucleases, which cut DNA strands at specific points. The gene is then copied billions fold in readiness for transfer to another organism. The uptake of foreign DNA or transgene by the animal or plant cell is called genetic transformations. There are several methods of introducing an isolated/foreign gene into the host plant. These include:

1. Micro-injections generally used in animals. The new DNA is injected with a very small sharp needle into the nucleus of a single cell (fertilized egg), which is then placed back in the female uterus to develop normally. Unfortunately there is a high rate of failure of using this technique. The cells infrequently take up and express the desired traits of the introduced DNA.
2. Biolistics is used in the genetic modification of plants and involves shooting new genes into the potential

host. Microscopic particles of gold or titanium are coated in the DNA sections which are to be introduced to the host. These are loaded into cartridges, similar to shotgun cartridges, which are then fired at the plant cells. The process relies on some of the microscopic particles entering this cell nuclei and their DNA coating combining with the plant chromosomes.

3. Vectors method has the potential to be used in both plant and animal situations. It involves a bacteria or virus carrying a new gene into a cell. Using a modification of what is already happening in nature e.g. *Agrobacterium tumefaciens* and *Agrobacterium rhizogenes.* These bacteria are usually found in the soil. These cause galls or hairy roots through introducing some of their own DNA into the plant. The *Agrobacterium* transfer the DNA as a plasmid, a small circular piece of DNA, which is separate to the main bacterial chromosome. Genetic engineering makes use of this natural transfer of DNA through replacing a section of the bacteria's own DNA with a gene which scientists would like to introduce to a new host.
4. Co-cultivation (Protoplast transformation) is also commonly used in plants. The cellulose in the plant wall is dissolved away using enzymes leaving a protoplast. DNA can then be added to the protoplast and then cultured on a growth media. This encourages the protoplast to regrow cell walls and eventually grow into a transgenic plant.
5. Other methods for the genetic transformation for the crop improvement include: electroporation, leaf disk transformation method, virus mediated transformation, pollen-mediated transformation, liposome- mediated transformation, etc.

The Scenario of Genetically Modified Crops in the World

The United States is the world's leading GM nation, both in terms of the area under cultivation and public acceptance of transgenic food. Crop varieties developed by genetic engineering were first introduced for commercial production in 1996. Worldwide, about 672 million acres of land are under cultivation, of which 25 per cent or 167.2 million acres consisted of GM crops. The U.S. farmers are by far the largest producers of genetically-modified (GM) crops using 105.7 million acres accounts for nearly two-thirds of all biotechnology crops planted globally, making up 40 per cent of the country's maize, 81 per cent of soya beans, 65 per cent of canola, or oilseed rape, and 73 per cent of cotton. Two virus-resistant crops, papaya and squash, are currently planted on small acreage (fewer than 7,000 acres) in the United States (Giannessi *et al.*, 2002). In the United States the three main GM crops under cultivation are varieties of corn, soybeans, and cotton. These crops had tremendous increase in terms of acreage over the years (Table 1). Other GM crops currently grown in the U.S. include canola, squash and papaya. Some commercially available GM crops, such as sugar beets, potatoes, and sweet corn, have yet to be widely adopted by farmers.

Argentina is the next largest producer, with 34.4 million acres, followed by Canada with 10.9 million acres, where its principle crop canola along with GM corn, soyabeans are widely grown; Brazil with 8.4 million acres, which has legalized the planting of GM soybeans; China with 6.9 million acres, where the acreage of GM cotton continues to increase; and South Africa with 1.0 million acres in 2003, where cotton is also the principle biotech crop (Fig. 1). Together, these six countries grew 99 per cent of the global GM crop area last year. Australia, Mexico, Romania, Bulgaria, Spain, Germany, Uruguay, Indonesia, the Philippines, India, Columbia, and Honduras also planted significant acreage in GM crops in 2003. Between 1996 and 2000, the area of genetically-modified (GM) crops increased from 4 to 44 million ha (James 2000).

Genetically-modified crops are very rare in Europe. Controversies about GM foods in the UK and several other European countries highlight the apparent differences that exist in public opinion as compared to United States (Gaskell *et al.,* 1999). In October 1998, France, Italy, Denmark, Greece and Luxembourg banded together declared 'no' to GM crops except Spain where a variety of herbicide resistant maize is being grown in any significant amount. Strict labelling laws and regulations have been introduced for GM food (DNA bar codes), and of course public opinion towards the technology is mainly negative. Now that the EU has agreed on regulations for labelling and tracking GM produce, the grounds for these countries' objections have largely been removed.

TABLE 1
Major Genetically-modified Crops of USA

Crop	*Total Acreage in 2001*	*Total Acreage in 2002*	*Total Acreage in 2003*	*Total Acreage in 2004*
Corn	75,800	79,000	79,066	81,100
Soybean	74,105	72,993	73,653	74,724
Cotton	15,499	14,151	13,924	13,947

Scenario of GM Crops in India

Government of India has not yet announced a policy on GM foods because on GM crops are still grown in India except *B.t.* cotton and no products are commercially available in supermarkets yet (38). India is, however, very supportive of transgenic crop research. Because, India is under tremendous pressure from the biotechnology industries to allow GM crops. These companies have the financial resources to mobilize scientific opinion as well as political support. More generally, attitudes towards GM crops are influenced by national pride, with opposition focused on the products of foreign multi-national companies, rather than homegrown technologies. It is highly likely that India will decide that the benefits of GM foods outweigh the risks because Indian agriculture will need to adopt drastic new measures to counteract the country's endemic poverty and feed its exploding population.

As the world's largest cotton growing nation, India could become a huge market for transgenic cotton. The government approved commercial planting of varieties engineered to produce the insecticidal *B.t.* protein in 2002, although in the first year only 0.5 per cent of the country's cotton was transgenic. However, *B.t.* cotton field trials revealed that such products are not suitable for Indian conditions. Besides cotton, genetic engineering experiments are being conducted on maize, mustard, sugarcane, sorghum, pigeon-pea, chick-pea, rice,

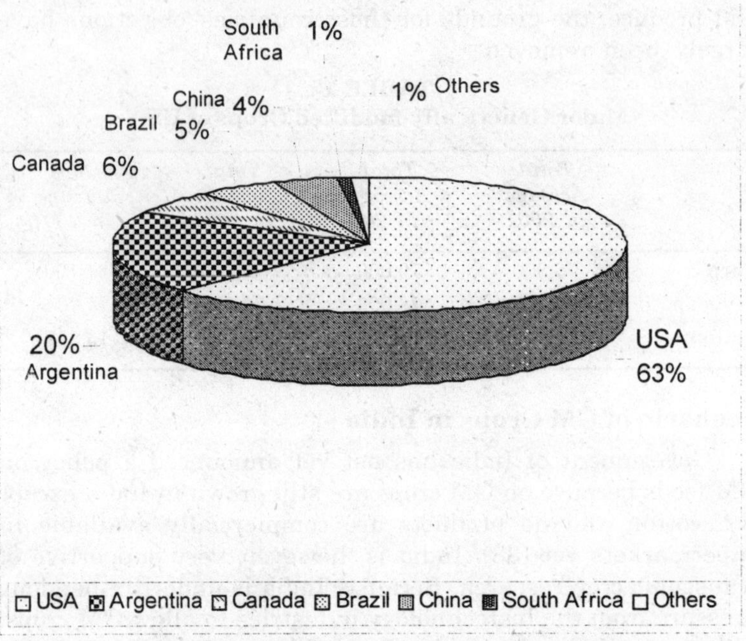

Fig. 1: Percent Global Land used for Genetically modified

tomato, brinjal, potato, banana, papaya, cauliflower, oilseeds, castor, soyabean and medicinal plants. Experiments are also underway on several species of fish. In fact, such is the desperation that scientists are trying to insert *B.t.* gene into any crop they can lay their hands on, not knowing whether this is desirable or not. India's government supports local research into protein rich potato, high yielding mustard, and drought as well as salt tolerant rice etc.

What is the Latest State of Affairs?

Genetically-modified food and agricultural biotechnology have generated considerable interest and controversy around the world. Because these foods contain genes derived from bacteria and viruses. These genetic changes are radically different from those resulting from traditional methods of breeding. Genetic engineering poses the greatest dangers of any technology because of its damaging effects are irreversible. It can never be reversed or cleaned up; genetic mistakes will be passed on to all future generations of a species. That's why; we must prevent problems before they occur. The precautionary approach is essential if we are to protect children, our future generations, and ourselves. We have to take action now, if we want to prevent a flood of genetically engineered foods from inundating the market and placing virtually everyone at risk.

The traditional crop breeding scientists are feeling several problems of the GM crops to the mankind and environment includes viz. introduction of new toxins and allergens in GM foods, serious health risks and diseases caused by unnatural foods, the creation of herbicide-resistant super weeds, loss of our crop bio-diversity and disturbance in ecological balance etc. Artificially induced characteristics and inevitable side effects will be passed on to all subsequent generations and to other related organisms. Once released, they can never be recalled or contained. The consequences of these are incalculable.

GM crops are widely used in food for both humans and animals, none of which needs to be labelled as transgenic. There is lot of controversies among the scientific community; some advocate the technology's benefits while others raise questions about environmental and food safety issues. Even though many scientists say that genetically-modified foods could cause serious damage to health and the environment. Yet, the sale of these foods is being permitted without proper assessment of the risks and adequate information to the public.

GM foods available, or about to appear, in U.S. markets include tomatoes, squash, yeast, corn, potatoes, and soyabeans (which are used in 60 per cent of all processed foods, such as bread, pasta, candies, ice-cream, pies, biscuits, margarine, meat products and vegetarian meat substitutes). Genetically-modified organisms are also used to produce cheeses and canola oil. But this is just the beginning. In a few years it may be almost impossible to find natural food. However, the assessment of genetically-modified plants/foods with enhanced nutritional properties should focused on the simultaneous characterization of inherent toxicological risks and nutritional benefits (Kuiper *et al.*, 2001). The food industry and government appear to be satisfied. They assume that these new foods are not substantially different from existing foods and pose no special risks.

Commercial Issues

The world population has crossed 6 billion people and is predicted to double in the next 50 years. Ensuring an adequate food supply for this booming population is going to be a major challenge in the years to come. So not only the taste, nutrients but increasing the yield is one of the greater thrust can be achieved by controlling environmental abiotic and biotic stresses. These crops have also shown the great promise for pest resistance, resistance to diseases, cold tolerance, drought tolerance/salinity tolerance, herbicide tolerance etc. and enhanced shelf life. As a consumer, one can enjoy all the benefits of GM Foods such as quality food (free of chemical residues). The farming community and the consumers are the primary beneficiaries. Farmers get better yields and their profits increase because of the reduction in cost of cultivation. Consumers will get good quality food that is free from chemical residues, which keeps them healthy. The technology opens up several new markets and thus helps the business community. Governments and health organizations reap benefits in the form of easier production of vaccine fruits and nutritious crops.

However, the chemical pesticide industry could lose business because of the reduced use of their products by farmers as they typically use many tons of chemical pesticides annually. Consumers do want to eat food that has been treated with pesticides because of potential health hazards, and run-off of agricultural wastes from excessive use of pesticides and fertilizers can poison the water supply and cause harm to the environment. Growing GM foods such as *B.t.* corn can help eliminate the application of chemical pesticides and reduce the cost of bringing a crop to market (Moellenbeck *et al.*, 2001).

Biotechnology is a big budget business with patent laws allowing private interests to develop and market unique products. All of those things previously mentioned are within the realms of possibility of genetic engineering and they will soon happen because of the commercial gain they represent. People generally have to pay for things they want whether it is a life saving banana or a pesticide free cabbage. The potential monetary benefits are huge. The value of the global market for genetically-modified crops is expected to be between 2 and 3 billion US dollars in the year 2000 increasing to 6 billion dollars in the year 2005. Supporters of GM technology argue that engineered crops like golden rice (vitamin A rich) or protein enhance potatoes can improve nutrition. The drought or salt-resistant varieties can also flourish in poor climatic conditions as the functionality of newer genes that are involved in plant adaptation to abiotic stresses, and can reduce starvation of that area.

The GE insect-repelling crops protect the environment by minimizing pesticide use. The final goal is to develop plants engineered for newer traits to help the human community produce food in adverse environmental conditions. Other benefits to business from using genetic technology are the ability to improve the quality, quantity and other desirable traits of products, remove undesirable traits and to hasten the introduction of superior genotypes to the commercial world. Currently conventional breeding can take decades to produce the right characteristics in a crop whereas the same result can be achieved in a matter of years using gene technology.

Due to evolution of new products and growing techniques, within another two decades, our knowledge about the functional importance of most plant genes will be nearly complete and the technology for producing GM crops will be far better than what is available now-a-days. This will dramatically increase the benefits we can get from GM crop that are engineered for drought tolerance, altered life cycle time, and nutrient balance; produce that can be easily processed; along with their enhanced flavor, etc. The technology has the potential to grow food in spaceships and stations with minimum resources. Technical improvements in the meat industry such as genetic manipulation and cloning of livestock will offer similar benefits by producing quality food in a cost-effective manner.

Transgenic animals will have better yields of meat, eggs, milk, improved animal health, increased resistance against pathogens and feed efficiency. DNA vaccines are already being developed for several animal diseases and these vaccines can help produce good quality meat more cheaply. The farming community and the consumers are the primary beneficiaries from the products of GM crops. Farmers get better yields and their profits increase because of the reduction in cost of cultivation. Consumers get good quality food that is free from chemical residues, which keeps them healthy. The technology opens up several new markets and thus helps the business community. Governments and health organizations reap benefits in the form of easier production of vaccine fruits and nutritious crops. The chemical pesticide industry could lose business because of the reduced use of their products by farmers. Engineering crop plants for pest resistance reduces the use of pesticides and storage costs, thereby reducing the cost of cultivation and environmental damage. However, farmers must be made aware of the potential dangers of pest resistance and herbicide tolerance traits to weed plants and their dangerous effect such as lateral genetic movements on the environment.

The most obvious economic advantage is higher yields due to the enhanced qualities of GM crops to defend themselves from adverse environmental factors, both abiotic and biotic. Growing GM crops that can withstand salinity and drought can reduce costs involved in soil reclamation. Other economic benefits are the general benefits received from the overall healthy living conditions of people as a result of using GMF. Potential disadvantages may be a loss of balance in the agriculture-dependent industry sectors because of reduced chemical inputs. Regulations such as patents and royalties may restrict easy access to research findings and thus affect the distribution of wealth.

Ethical Issues

The concepts of ethics differ from country to country according to culture and condition. These also change with time due to shifting of perception of values. An example, *B.t.* cotton was not ethically wrong, but some find it as inherent problems in the use of gene across species. They consider, this is violation of natural organisms' intrinsic values. Some characterized this tampering with nature by mixing genes among species as "playing with God's creations", of course ethically unacceptable to interfere with nature. The issue leads the questioning of many long held traditions and beliefs. And these need to be answered by scientific community before we have an adequate basis of knowledge for reaching and ethical conclusion.

Levidow and Carr (2000) have been suggested that all environmental controversies at root involves disputes about fundamental principles, however ethical issues are sidelined for GM crops. While people will be put off or even outraged by such possibilities, technologists point out that although there may be an ethical dilemma which is likely to be debated emotively, the chemical structure of DNA is the same whether a human, a tree or an amoeba. It is only the sequence of the nucleotides within the DNA that determines the genetic makeup of the organism. New ethical questions have arisen due to intervene the structure of genome (Polkinghorne, 2000).

In GM crops, genes are transferred from any organism to another. Transferring animal genes into plants raises important ethical issues for vegetarians and religious groups. It may also involve animal experiments, which are unacceptable to many people. For example if vegetarian people come to know that their food from genetically-modified crops contains DNA copied from a pig gene:

- What will be their mental status?
- If it were to contain copies of a human gene does this mean that the person eating it is a cannibal?

So, there may be a lot of objections for consuming animal genes in plants and *vice versa*.

Further the lack of labelling violates the right to know what is present in our foods, we are eating. Even if GM foods are absolutely safe, the consumer has a right to know its ingredients, its impact on health and environmental. Under present regulations manufacturers are already introducing genetically-modified ingredients into many processed foods without informing consumers. The government is ignoring the wishes of the public. Surveys consistently find that 85-90 per cent of consumers want clear labelling of all genetically engineered foods. Food labels must be designed to clearly convey accurate information about the product in simple language that everyone can understand. This may be the greatest challenge faced be a new food labelling policy: how to educate and inform the public without damaging the public trust and causing alarm or fear of GM food products.

In January 2000, an international trade agreement for labelling GM foods was established (Helmuth, 2000). More than 130 countries, including the US, the world's largest producer of GM foods, signed the agreement. The policy states that exporters must be required to label all GM foods and that importing countries have the right to judge for themselves the potential risks and reject GM foods, if they so choose. This new agreement may spur the U.S. government to resolve the domestic food-labelling dilemma more rapidly.

Intellectual property rights are another ethical issues likely to be an element seeking the debate on GM foods, with an impact on the rights of farmers, as domination of world food production will be by a few companies. If food production is monopolized, the future of that supply becomes dependent on the decisions of a few companies and the viability of their seed stocks.

Other problems include serious health hazards. Many kids in the US and Europe have developed allergies against the GM foods. There is a growing concern that introducing foreign genes into food plants may have an unexpected and negative impact on human health. As we cannot rule out the possibility that transgene may create a new allergen or cause an allergic reaction in susceptible individuals. An example a gene from Brazil nuts into soybeans was discarded because of the fear of causing unexpected allergic reactions. Extensive testing of GM foods is required to avoid the possibility of harm to consumers with food allergies. Investigations of Nordlee *et al.* (1996) revealed that GM potatoes, spliced with DNA from the snowdrop plant and a viral promoter (CaMV), the resulting plant was poisonous to mammals (rats), damaging their vital organs, the stomach lining and immune system. CaMV is a para-retrovirus. All transgenic crops containing CaMV 35S or similar promoters which are recombinogenic should not be permitted for commercial production or open field trials. All products derived from such crops containing transgenic DNA should also be immediately withdrawn from sale and from use for human consumption or animal feed.

Engineered Crops Produce Their Own Pesticides

Hence, this will promote the more rapid appearance of resistant insects and lead to excessive destruction of useful insects and soil organisms, thus seriously perturbing the ecosystem. The most commonly used insecticide is a toxin from the bacterium *Bacillus thuringiensis* (*B.t.*). This modified transgene is not always likely to express (turned on) throughout the whole plant. It may be confined to the leaves of the plant. However, if the gene is expressed in pollen then there is a chance that the insecticide may harm beneficial insects that eat the pollen. Studies have shown that GM products such as pollen from *B.t.* corn caused high mortality rates of beneficial insects, most notably the monarch butterfly larvae (Losey *et al.*, 1999). *B.t.* crops are also indirectly affect other beneficial insects. Swiss researchers have shown that green lacewings, beneficial predatory insects, suffered a higher death rate and delayed development when fed European corn borers which had eaten *B.t.* corn compared with lacewings fed borers given non-*B.t.* corn (Hilbeck *et al.,* 1998). Further, Insecticides expressed by GM plants may not only harm the target pest species. They could also pose a danger to the pest insect's natural enemies when they ingest the pest insect, along with traces of the toxic pesticide.

Environmental Issues

The studies have revealed the potential impacts of GM crops/products on the environment (Letourneau *et al.*, 2002). Altering the genes in an organism has a great many effects on the characteristics of the crops. Introducing such a crop into the environment carries the great risk of altering the balance of an ecosystem. Gene escape or the gene flow i.e. the exchange of genetic information among individuals or species, from GM crops is often raised the issue of biosafety and environmental concerns (Quist and Chapela, 2001). The evolution of superweeds and loss of natural diversity is an important concern. In considering evolutionary concerns, it is important to mention the effects of transgenics on other organisms in the ecosystems. Anti-GM entities consider it to be the biggest uncontrolled biological experiment going on in the world today. Although proof of serious harm to humans, animals and plants has yet to be definitively proven. But the opponents fear that the environment could be damaged through accidental cross-pollination of GM products with natural plants.

According to Rissler and Mellon (1996) the Ecological Risks of genetically-modified crops pose six kinds of potential risks.

1. The GM crops themselves could become weed/super weed with undesirable effects.
2. From the GM crops new genes move to wild plants, modifying their original nature, of the which could then become weeds.
3. GM crops may produce viruses could facilitate the creation of new, more virulent or more widely spread viruses.
4. These plants express potentially toxic substances could present risks to other organisms, which use them as food like birds or deer.
5. These crops may have effects that flow through an ecosystem in such a ways that are difficult to predict.
6. Finally, these crops might threaten centers of crop diversity.

Other potential harmful impact being includes transfer of antibiotic resistance markers, potential environmental impact: undesirable effects on other organisms (e.g., soil microbes), loss of flora and fauna biodiversity. Any new organisms may be more competitive than native flora and fauna in the native species own ecological niche. It is everybody's responsibility to ensure native flora and fauna are sustained for future generations. More than 50 per cent of the crops developed by biotechnology companies have been engineered to be resistant to herbicides. Use of herbicide-resistant crops will lead to a threefold increase in the use of herbicides, resulting in even greater pollution of our food and water with toxic agrochemicals

(Goldberg, 1992). So, it will further lead to soil sterility and increased pollution of food and water supply. Insecticides and other products released by GM plants are released into the soil. In New Zealand, scientists are monitoring the effect that *B.t.* transgenic potatoes have on soil. So far, only minor and transient effects have been found, though the research is still going on.

Dangers of Superweeds

The GM crops engineered for herbicide tolerance and weeds can cross-breed, resulting in the transfer of the herbicide resistance genes from the crops into the weeds or that they could accidentally breed with wild plants or other crops. More conceivable threats are that the resulted crops could become insidious superweeds. In other words, we can say that genetically polluting the environment. These "superweeds" would then be herbicide tolerant as well. Other introduced genes may cross over into non-modified crops planted next to GM crops. This could be a potentially serious problem if "pharm" crops, engineered to produce pharmaceutical drugs, accidentally cross breed with food varieties (or seeds become mixed up). Studies have revealed that that genetically-modified *B.t.* endotoxin remains in the soil at least up to 18 months (Marc Lappe and Britt Bailey) and can be transported to wild plants creating superweeds growing nearby in just one generation (Mikkelsen, 1996). A U.S. study showed the superweed resistant to glufosinate to be just as fertile as non-polluted weeds. Another study showed 20 times more genetic leakage with GM plants — or a dramatic increase in the flow of genes to outside species. Also in a U.K. study by the National Institute of Agricultural Botany, it was confirmed that superweeds could grow nearby in just one generation. Scientists suspect that Monsanto's wheat will hybridize with goat grass, creating an invulnerable superweed. An experiment in France showed a GM canola plant could transfer genes to wild radishes. Similarly, report published in New Scientists, an Alberta Canada farmer began planting three fields of different GM canola seeds in 1997 and by 1999 produced not one, but three different mutant weeds - respectively resistant to three common herbicides. In effect genetic materials migrated to the weeds they were meant to control. This could pose a high cost to farmers and threaten the ecosystem.

Genes are exchanged between plants via pollen. One of the main difficulties which farmers will encounter when growing GM crops is that there is no way to contain pollen movement. In the case of oilseed rape, researchers have found that its pollen can travel up to 4 km and can escape from fields even when they are surrounded by barrier crops to prevent this. But the evidence shows that this is clearly not enough to protect farmers and consumers from GM contamination. There have already been several serious incidents of GM contamination, despite the fact that GM crops are only grown by a minority of farmers worldwide.

One way to ensure that non-target species will not receive introduced genes from GM plants are to create GM plants that are male sterile (do not produce pollen) or to modify the GM plant so that the pollen does not contain the introduced gene (Daniell, 1999). Cross-pollination would not occur, and if harmless insects such as monarch caterpillars were to eat pollen from GM plants, the caterpillars would survive. Another possible solution is to create buffer zones around fields of GM crops. For example, non-GM corn would be planted to surround a field of *B.t.* GM corn, and the non-GM corn would not be harvested. Beneficial or harmless insects would have a refuge in the non-GM corn, and insect pests could be allowed to destroy the non-GM corn and would not develop resistance to *B.t.* pesticides. Gene transfer to weeds and other crops would not occur because the wind-blown pollen would not travel beyond the buffer zone. Estimates of the necessary width of buffer zones range from 6 meters to 30 meters or more. This planting method may not be feasible if too much acreage is required for the buffer zones.

Conclusions

As we move into the 21st Century and world population increases towards 9 billion, cereal grain production will need to increase by one billion tones (Borlaug, 2001). Against a background of static or declining area of land available for cultivation, GM crops have the potential to solve many of the world's hunger and malnutrition problems, and to help protect and preserve the environment by reducing reliance upon chemical pesticides and herbicides. There is also the need for this technology to increase crop yield, improve nutritional quality of food and reduce crop losses. Yet there are many challenges ahead for governments, especially in the areas of safety testing, regulation, international policy and food labeling. Many people feel that genetic engineering is the unavoidable wave of the future and that we cannot afford to ignore a technology that has such enormous potential benefits. However, there are many controversial issues associated with the introduction of GM crops, but we must proceed with caution to avoid causing unintended harm to human health and the environment as a result of our enthusiasm for this powerful technology.

REFERENCES

Betz, F.S., Hammond, B.G., and Fuchs, R.L. (2000). Safety and advantages of *Bacillus thuringiensis*-protected plants to control insect pests. *Regul. Toxicol. Pharmacol*. 32: 156-173.

Carr, S. and Levidow, L. (2000). Exploring the links between Science, risk, uncertainity and ethics in regulatory controversies about genetically-modified crops. *J. of Agriculture and Environmental Ethics* 12: 29-39.

Daniell H. (1999). New tools for chloroplast genetic engineering. *Nature Biotechnology* 17(9): 855-856.

Gaskell G., Bauer M.W., Durant J. and Allum N.C. (1999). Worlds Apart? The Reception of Genetically Modified Foods in Europe and the U.S., *Science,* 16 July 1999, Vol. 285: 384-387.

Giannessi L.P., Silvers C.S., Sankula S., and Carpenter J.F. (2002). *Plant Biotechnology: Current and Potential Impact for Improving Pest Management in US. Agriculture*. Washington, D.C.: National Center for Food & Agricultural Policy. http://www.ncfap.org.

Goldberg, R.J. (1992). Environmental concerns with the development of herbicide-tolerant plants. *Weed Technology*, 6: 647-652.

Gressel, J. (1999). Tandem constructs: preventing the rise of superweeds. *Trends in Biotech*., 17: 361-366.

Helmuth, L. (2000). Biotechnology: Both sides claim victory. *Science* 287: 782-783

Hilbeck, A., Baumgartner, M., Fried, P. and Bigler, F. (1998). Effects of transgenic *Bacillus thuringiensis* corn-fed prey on mortality and development time of immature *Chrysoperla carnea* (Neuroptera: Chrysopidae). *Environmental Entomology* 27: 480-487.

James, C. (2000). Global review of commercialized transgenic crops: 2000. *ISAAA Briefs.*

Kuiper, H.A., Kleter, G.A., Noteborn, H.P.J.M. and Kok, E.J. (2001). Assessment of the food safety issues related to genetically-modified foods. *The Plant Journal*, 27(6): 503-528.

Letourneau, D. and Burrows, B. (eds.) (2002). *Genetically Engineered Organisms: Assessing Environmental and Human Health Effects.* Boca Raton, Fla.: CRC Press, 438 p.

Levidow, L. and Carr, S. (2000). Unsound science? Transatlantic regulatory disputes over GM crops. *International Journal of Biotechnology* 2(1-3): 257-273.

Losey, J.E., Rayor, L.S., and Carter, M.E. (1999). Transgenic pollen harms monarch larvae. *Nature,* 399: 214.

Mikkelsen, T.R., Jensen, J. and Jorgensen, R.B. (1996). Inheritance of oilseed rape *(Brassica napus)* RAiPD markers in a backcross progeny with *Brassica campestris. Theor. Appl. Genet.* 92: 492-497.

Moellenbeck, D.J., Peters, M.L., Bing, J.W. *et al.* (2001). Insecticidal proteins from *Bacillus thuringiensis* protected corn from corn rootworms. *Nature Biotechnology,* 19(7): 668-672.

Nordlee, J.A., Taylor, S.L., Townsend, J.A., Thomas, L.A. and Bush, R.K. (1996). Identification of a Brazil nut allergen in transgenic soybeans. *N Engl J Med.*, 334: 688-692.

Polkinghorne, J.C. (2000). Ethical issues in biotechnology. *Tibtech.*, 18: 8-10.

Quist, D. and Chapela, D. (2001). Transgenic DNA introgressed into traditional maize landraces in Oaxaca, Mexico. *Nature*, 29: 414 (6863): 541-543.

Rissler, J. and Mellon, M. (1996). *The Ecological Risks of Engineered Crops.* Cambridge, Mass.: MIT Press, 168 p.

CHAPTER 6

Assessment of Medicinal Plants Growing in Shakumbhari Devi Hills of Shivalik Range

Piyush Kumar Patel* and Manjul Dhiman*

ABSTRACT

A local survey of Shakumbhari Devi hills was carried out periodically. The area is known to have a rich biodiversity. A number of medicinal plants are growing in this area. Nearly hundred plants of medicinal and ethonobotanical interests are collected from the area and are documented. Surprisingly, exploitation of these plants is occurring to a great extent.

Harvesting of different parts of drug plants is carried out at time to time by commercial users. Tonnes of plant products are traded to nearby areas. It is also observed that overgrazing is one of main causes of loss of biodiversity in this area by cattles of Van Gujjars.

Introduction

In India, initially medicine men were always religious people. Every village used to have a small medicinal garden which was used to provide them raw material used by village *vaidyas* or *hakims* (Husain, 1982). Ethno-botanical studies lay emphasis on the floristic diversity preserved in primeval forests in the form of sacred grooves, largely maintained by tribals because of rituals beliefs or taboos (Arora, 1997).

Shivalik hills of Saharanpur, Haridwar and Dehradun districts is recognized as an inhabitant of numerous drug and medicinal

*Department of Botany, K.L.D.A.V. (P.G.) College, Roorkee, Uttarakhand.

plants. Areas of Shakumbhari hills in Saharanpur is occupied by saints and *ojhas*. Villagers residing in these areas consult local medicine men and saints for treating their ailments.

Shakumbhari Devi is situated in Shivalik range towards 40 km north of district Saharanpur. The district lies between 77° - 15′ to 77° - 55′ east latitude and 29° - 24′ to 30° - 30′ north longitude. The altitude above sea level is about 340 metre. The maximum height of Shakumbhari hills is 909 metres. Area has stony and gravel soil. Shakumbhari hills have many seasonal drains.

Shakumbhari Devi hills are exploited by several traders of medicinal plants. The area is very rich in medicinal plant wealth.

Ahuja (1965) listed 92 plants species of medicinal importance in the adjoining area of Saharanpur including Haridwar region. Similarly Uniyal (1965, 1977) and Bhargava (1982) explored medicinal plants from neighbouring areas of Uttarakhand and Saharanpur. Information on medicinal plants of these areas is also updated by Dhiman (1997, 2001, 2002, 2003) and Dhiman and Kaushik (1999) and Kaushik and Dhiman (2000).

It is estimated that annual overall demand for vegetative drugs in the past few decades is rising by nearly 7 per cent in organized sectors of trade (Gupta and Sethi, 1982). Pharmaceutical industries give greater emphasis and attention to chemical and biochemical parameters. Plants exploration under taken by industries in several parts of world provides sufficient testimony in this direction (Gupta, 1977).

Observation

Present report describes great medicinal plant wealth of Shakumbhari Devi hills. We have surveyed the area and compiled a list of these medicinal plants and their uses as described by local people, tribals, van gujjars, saints etc. residing in these areas (See Table 1).

Results and Discussion

It is observed that nearly hundreds of medicinal plants occur in this area. These plants are commonly used in various ailments. There is a serious threat to biodiversity of this area as traders are collecting tonnes of plants and their parts from this area. The hilly area is in habitant of van gujjars ,the people depending upon natural resources and cattles for milk. Overgrazing by these cattles is also causing loss to biodiversity of this area. There is a need to protect and conserve this area.

REFERENCES

Ahuja, B.S. 1965. *Medicinal Plants of Saharanpur*. Bishen Singh Mahendrapal Singh, Dehradun.

Arora, R.K. 1997. Ethnobotany and its role in the conservation and use of plant genetic resources in India. *Ethnobotany*, 9: 6-15.

Bhargav, S. 1982. Some Medicinal Plants of Distt. Saharanpur. M. Phil. Thesis, Meerut University, Meerut.

Dhiman, A.K. 1997, A survey of Medicinal Plants of Haridwar and Adjoining Area *vis-a-vis* the raw plant drugs being sold in local market. Ph.D. Thesis. Gurukul Kangri University, Haridwar.

Dhiman, A.K. 2001. Medicinal plant wealth of Haridwar and adjacent area. *Sachitra Ayurved*, 55: 914-925.

Dhiman, A.K. 2002. Common bazar medicines of Haridwar (Uttaranchal). *Sachitra Ayurved*, 55: 613-619.

Dhiman, A.K. 2003. Sacred plants of district Haridwar (Uttranchal) and their medicinal uses. *Ad. Plant Sci.*, 16: 377-384.

Dhiman, A.K. and Kaushik, P. 1999. "Avenue trees for conservation of environment at Haridwar". In: D.R. Khanna, and Gautam, A. (Eds), *Sustainable Ecosystem and Environment*. Action for Sustainable Development and Awareness (ASEA), Rishikesh.

Gupta, R. 1977. Medicinal plant components of Indian forest. *Indian Fmg*. 25: 95-98.

Gupta, R. and Sethi, K.L. 1982. "Conservation of medicinal plant resources in the Himalayan region." In: Jain, S.K and Mehra, K.L. (Eds), *Conservation of Tropical Plant Resources*. Botanical Survey of India, pp. 101-109.

Husain, A. 1982. "Conservation of genetic resources of medicinal plants in India". In: Jain, S.K. and Mehra, K.L. (Eds). *Conservation of Tropical Plant Resources*. Botanical Survey of India, pp. 110-117.

Kaushik, P. and Dhiman, A.K. 2000. *Medicinal Plants and Raw Drugs of India*. Bishen Singh Mahendra Pal Singh, Dehradun.

Uniyal, M.R. 1965. *Saharanpur Van Khand Ki Vanoushdhi*, 1962-63 (in Hindi). *Sachitra Ayurved* 18: 39-44.

Uniyal, M.R. 1977. *Uttrakhand Vanoushdhi Darshika*. Central Council of Research in Indian Medicine and Homeopathy, Delhi.

TABLE 1
List of Commonly Found Medicinal Plants of Interest in Shakumbhari Hills.

Botanical Name	*English Name*	*Vernacular Name*	*Uses for diseases*
Abutilon indicum Linn.	Indian mallow	Atibala	Febrifugic, anticancer, anti-inflammatory astringent.
Acacia catechu Willd.	Cutch tree	Katha/Khair	Antiseptic, astringent, anti-inflammatory used in skin diseases.
Acacia concinna D.C.	Soapnut	Shikakai	Purgative used in jaundic, cathartic, chronic cough.
Acacia nilotica (L) Willd.	Allergic black wood	Babool	Astringent, demulcent used in bronchitis.
Achyranthes aspera Linn.	Prickly chaff flower	Chirchita/Latjira	Used in cough, cold, snake bite, hydrophobia.
Adhatoda vasica Nees.	Malabar nut	Arusa/Vasaka	Used in cough, chronic bronchitis, asthma, diarrhoea, dysentery.
Adiantum capillus veneris Linn.	Maidenhair Fern	Hansraj	Expectorent used in cough, blood purification, ulcer, fever, liver disorders.
Aegle marmelos Corr.	Wood apple	Bael	Used in diabetes, cholera, nightfever, cromps.
Albizzia lebbek (Linn) Benth.	East Indian walnut	Siris	Astringent used in cough, asthma, scabies, anticancer.
Allium cepa Linn.	Onion	Pyaj	Stimulent, diuretic used in dysentery, piles, bronchitis.
Allium sativum Linn.	Garlic	Lahsun	Used in malaria fever, epilepsy, flatulence, ulcer.

Aloe barbadensis Mill.	Barbados Aloe.	Gwarpata/Ghrita kumari	Purgative, anthelmintic used in liver and spleen disease, colic pains.
Amaranthus spinosus Linn.	Spiny pig weed	Kateli chaulai	Used in fever, cough diarrhoea, appetizer.
Asparagus adscendens Roxb.	Musale	Safed musli	Used in diabetes, seminal debility, diarrhoea, general debility.
Asparagus racemosus Willd.	Asparagus	Satawar	Used in leucorrhoea, general debility, agalactia, hysteria.
Azadirachta indica A. Juss.	Margosa tree	Neem	Blood purifier used in malaria fever, antiulcer, fungal infection, syphilis.
Bacopa monieri Linn.	Bacopa	Neer brahmi	Blood purifier, memory enhancer used in dyspepsia, cough, fever.
Bambusa arundinacea Willd.	Bamboo	Bans	Used in bronchitis, cut wounds, fever, asthma, leprosy.
Berberis aristata DC.	Indian berberry	Daruhaldi	Stomachic, blood purifier, antipyretic.
Barleria prionitis Linn.	Barleria	Peela vasa	Expectorent, diphoretic, diuretic used in abadominal disorder.
Bauhinia variegata Linn.	Mountains ebony	Red kachnar	Used in diabetes. ulcer, obesity, asthma.
Bauhinia vahlii Weight.	Bauhinia	Maljan	Demulcent, antiseptic, antimicrobial.
Bombax ceiba Linn.	Silk cotton tree	Semul	Demulcent used in small pox, bleeding gum and kidney ulcers.
Butea monosperma Taub.	Forest flame	Tesu/Palash	Diuretic used in snake bite, caugh, tumour, fever.
Caesalpina crista Linn.	Fever nut	Karanju	Anthelmintic used in liver disorder, swelling.

(*contd.*)

Botanical Name	English Name	Vernacular Name	Uses for diseases
Calotropis procera Ait.	Akund	Madar/Aak	Used in cold, cough, asthma, dysentery.
Cannabis sativa Linn.	Indian hemp	Bhang	Analgesic, sedative, intoxicant.
Carissa carandas Linn.	Karandas	Karonda	Anthelmintic, hemiplegia.
Cassia absus Linn.	Four leaved cassia	Ban kulthi/chakshu	Used in leucoderma, leprosy, conjunctivitis, cataract.
Cassia fistula Linn.	Riding pipe tree	Amaltash	Blood purifier, emetic used in diabetes, cancer and epilepsy.
Cassia occidentalis Linn.	Negro coffee	Kasonda	Purgative, diuretic, febrifugic aphrodisiac.
Catharanthus roseus Linn.	Red periwinkle	Sadabahar	Diuretic, antitumour used in leucomia, diabetes.
Cedrela toona Roxb.	Red cedar	Tuna	Astringent used in, chronic, dysentery, blood sugar, ulcer.
Celastrus paniculatus Willd.	Celastrus	Malkangni	Antiseptic, laxative used in bronchitis, rheumatism, colic pains, leprosy, paralysis.
Chenopodium album Linn.	Lamb's guarters	Bathua	Anthelmintic, blood purifier used in ulcer piles, hepatitis.
Coccinia indica Wanda	Ivy gourd	Kanduri	Antimicrobial used in diabetes, gonorrhoea, syphilis, colitis.
Cordia dichotoma Forst f.	Indian cherry	Lasora	Antiseptic, antimicrobial used in snake bite, dyspepsia, fever.
Crateva nurvala Buch.	Crateva	Barna	Antipyretic, epidydimitis, demulcent.

Crinum asiaticum Linn.	Crinum	Sudarsna/Pindar	Expectorant anti-inflammatory, laxative, emetic used in swelling, cough, cold, vomiting.
Cuscuta reflexa Roxb.	Dodder	Amarbel	Purgative, anthelmintic, anodyne used in eczema, paralysis.
Cynodon dactylon Linn.	Bermuda grass	Doob ghas	Astringent, diuretic, epilepsy.
Dalbergia sisso Roxb.ex.DC.	Sissu	Shisham	Antiseptic, astringent, anti-bacterial used in leprosy, gonorrhoea.
Datura sp.	Thorne apple	Dhatura	Sedative, antiseptic, depressant used in asthma.
Desmodium gangeticum DC.	Desmodium	Sarivan/salparni	Astringent, diuretic used in chronic fever, snake bite.
Diospyros melanoxylon Roxb.	Coromandel ebony	Tendu	Astringent, diuretic carminative used in diarrhoea.
Emblica officinalis Gaertn.	Goosebery	Amla	Refrigerant used in epilepsy, gonorrhoea, anthrax.
Ervatamic coronaria (Jacq) syn.	Moon beam	Chandni	Refrigerant, vermicide, anodyne used in eye trouble.
Eucalyptus globules Labill.	Blue gum tree	Eucalyptus	Germicidal, disinfectant used in asthma, bronchits.
Eulophia pratensis Lindl.	Eulophia	Satavari	Anthelmintic used in bronchitis and blood diseases.
Euphorbia hirta Linn	Milk weed	Dudhi	Vermicidal used in leucoderma, dysentery, colic pain.

(contd.)

Botanical Name	English Name	Vernacular Name	Uses for diseases
Euphorbia nerifolia Linn.	Milk hedge	Sehund	Purgative, antiseptic used in colic pain, oedema, skin disease, scorpion bite.
Ficus benghalensis Linn.	Banyan tree	Vat Vraksha	Used in rheumatic pains, genital disorder, diarrhoea.
Ficus glomerata Roxb.	Lusterfig	Gular	Antimicrobial used in dysentery, diabetes, diarrhoea.
Ficus religiosa Linn.	Peepal	Pipal	Astringent, antimicrobial, purgative used in gonorrhoea.
Glycyrrhiza glabra Linn.	Licorice	Mulethi	Used in cough, peptic ulcer, epilepsy.
Helicteres isora Linn.	East Indian screw tree		Marorphali Used in skin diseases, stomach complaints, colic pains.
Holarrhena antidysenterica Linn	Tellicherry bark	Kura	Blood purifier used in diarrhoea, bronchopneumonia, chronic fever, anaemia, colites
Lawsonia inermis Linn.	Henna	Mehndi	Antiseptic used in jaundice, cooling, leprosy, spleen enlargement
Melia azedarach Linn.	Parsian lillac	Bakain	Anthelmintic diuretic used in spleen enlargement, leprosy.
Mimosa pudica Linn.	Tough me not	Lajwanti	Used in piles, fistula, urinary camplaints, germicidal.
Mirabilis jalapa Linn.	Four O'clock plant	Gulabans	Antiseptic, aphrodisiac, purgative used in stomach ailments.

Momordica dioca Roxb.	Wild gourd	Kakora	Antimicrobial used in asthma, cough, fever, leprosy.
Morus acedosa Griff.	Indian Mulberry	Tut	Anthelmintic, purgative, astringents, carminative.
Morus alba Linn.	White Mulbury	Shahtoot	Antiseptic, purgative, vermifuge, diaphoretic.
Mucuna prurita Linn. DC.	Cow hedge	Kawanch	Aphrodisiac, nervine tonic, anthelmintic, stimulant, purgative.
Murraya koenigii Linn.	Curry leaf plant	Karipatta	Stimulent used in insect bite, dysentery, indigestion.
Nyctanthes arbor-tristis Linn.	Night jasmine	Harsinghar	Diuretic, antibilious used in rheumatism, fever, sciatica.
Phoenix acaulis Roxb.	Wild date	Jangli khajoor	Used as aperient.
Prosopis spicigeral Druce	Sponge tree	Chookar	Antimicrobial, laxative used in rheumatism.
Psidium guajava Linn.	Guava	Amrud	Antiseptic, astringent, diuretic, diarrhoea.
Pueraria tuberosa DC.	Saral	Vidari kand	Demulcent, refrigerant, spermopoeitic,l anti-inflammatory, oestrogenic.
Ricinus communis Linn.	Caster	Arandi	Purgative, anti-inflammatory.
Santaloides minus Schellenb.	Santaloides	Vidhara	Used in rhemastism, syphilis, diabetes, ulcer.
Shorea robusta Gaertn.	Sal	Sal	Astringent, aphrodisiac used in dysentery.

(contd.)

Table 1 *(concld.)*

Botanical Name	English Name	Vernacular. Name	Uses for diseases
Sida cordifolia Linn.	Country mallow	Bala/Kharenti	Energy enhancer, beneficial in spermatorrhea, leucorrhoea.
Solanum nigrum Linn.	Black nightshade	Makoi	Antidysenteric, diuretic used in liver disorders.
Solanum surattense Burn.	Yellow nightshade	Kateli	Used in asthma, chronic bronchitis and rheumatism.
Strychnos nux-vomica Linn.	Nuxvomica	Kuchla	Used in ulcers, dyspepsia, epilepsy
Swertia chirayita Roxb.	Chirayata	Chirayata	Used in leprosy, leucoderma, asthma.
Syzygium cumini (L) Skeels.	Black plum	Jamun	Astringent used in jaundice, diabetes, diarrhoea, dysentery.
Tamarindus indica Linn.	Tamarind	Imli	Carminative used in liver ailments.
Tectona grandis Linn.	Teak	Sagaun	Astringent diuretic used in bronchitis, urinary problems.
Terminalia arjuna Roxb.	Arjun	Arjun	Cardiotonic, astringent, febrifugic, antiseptic.
Terminalia bellirica Roxb.	Myrobalan	Bahera	Diuretic, purgative, hair tonic used in rheumatic swelling.
Terminalia chebula Retz.	Chebulic myrobalans	Harad	Diuretic, cardiotonic used in stomach ailments.
Terminalia tomentosa Kurz	Spinous kinotree	Saj/Asan	Cardiotonic used in diabetes, cough.
Tinospora cordifolia Willd.	Tinospora	Giloy	Antipasmodic, antiinflammatry used in

			general debility.
Tribulus terrestris Linn.	Puncture vine	Gokhru	Urine trouble, diuretic, dyspepsia, hepatitis.
Urginea indica Kunth	Indian squill	Jangli Pyaj	Cardiotonic, expectorant, diuretic used in rheumatism.
Ziziphus jujuba Mill.	Wild plum	Ban ber	Anodyne used in nausea.

CHAPTER 7

Bio-Treatment of Organic Waste

Archana Mehta*, Shruti Shukla* and Abhinav Mahta[+]

Introduction

Microbial decomposition uses micro-organisms (bacteria and fungi) to biologically degrade hydrocarbon-contaminated waste into non-toxic residues. The objective of bio-treatment is to accelerate the natural decomposition process by controlling oxygen, temperature, moisture, and nutrient parameters. Land application is a form of bio-remediation. This focuses on forms of bio-remediation technology that take place in more intensively managed programs, such as composting, vermiculture, and bioreactors.

For centuries, one of the most eternal problems facing man-kind has been the dilemma of how to dispose of our garbage. Today, with global populations expected to climb into the tens of billions within the next century, never before in history has the issue of garbage disposal been more pressing. Our society must find new ways not only to reduce, reuse, and recycle our materials, but to detoxify our wastewater, soils, and environment as well. Widespread use of Effective Micro-organisms (EM) to detoxify our landfills, decontaminate our environment, and promote highly sustainable, closed-cycle agricultural and organic waste treatment methods worldwide.

Bio-treatment of soil organic material is a key process for the global carbon cycle. Decomposition processes may be limited by available resources, predominantly by carbon and nitrogen. Thus, increasing N input to terrestrial ecosystems by enhanced

---

* Department of Botany and [+]Department of Pharmaceutical Sciences, Dr. H. S. Gour University, Saugor - 470 003, India.

anthropogenic N deposition may strongly affect decomposition and thereby carbon release from soils. The rate of decomposition is dependant on the size and species of the material and the geographic location (Maser *et al.,* 1979).

Role of Micro-organisms in Decomposition

Microorganisms such as bacteria, fungi and actinomycetes account for most of the decomposition, as well as the rise in temperature that occurs in the compost process. Some microbes require oxygen to function, others do not. Those requiring oxygen are preferred in composting. Also, different micro-organisms thrive in different temperature ranges. The goal in constructing and managing compost is to create an environment suitable for the desired micro-organisms (Kimmins, 1991).

In the process of decomposting, micro-organisms break down organic matter and produce carbon dioxide, water, heat, and humus, the relatively stable organic end product. Under optimal conditions, composting proceeds through three phases: (1) the mesophilic, or moderate-temperature phase, which lasts for a couple of days, (2) the thermophilic, or high-temperature phase, which can last from a few days to several months, and finally, (3) a several-month cooling and maturation phase.

Different communities of micro-organisms predominate during the various composting phases. Initial decomposition is carried out by mesophilic micro-organisms, which rapidly break down the soluble, readily degradable compounds. The heat they produce causes the compost temperature to rapidly rise (Adams *et al.,* 1995). As the temperature rises above about 40°C, the mesophilic micro-organisms become less competitive and are replaced by others that are thermophilic, or heat-loving. At temperatures of 55°C and above, many micro-organisms that are human or plant pathogens are destroyed. Because temperatures over about 65°C kill many forms of microbes and limit the rate of decomposition, compost managers use aeration and mixing to keep the temperature below this point. During the thermophilic phase, high temperatures accelerate the breakdown of proteins, fats, and complex carbohydrates like cellulose and hemicellulose, the major structural molecules in plants. As the supply of these high-energy compounds becomes exhausted, the compost temperature gradually decreases and mesophilic micro-organisms once again take over for the final phase of "curing" or maturation of the remaining organic matter.

Bacteria

Bacteria are the smallest living organisms and the most numerous in compost; they make up 80 to 90% of the billions of micro-organisms typically found in a gram of compost. Bacteria are responsible for most of the decomposition and heat generation in compost. They are the most nutritionally diverse group of compost organisms, using a broad range of enzymes to chemically break down a variety of organic materials. Bacteria are single-celled and structured as rod-shaped bacilli, sphere-shaped cocci or spiral-shaped spirilla. Many are motile, meaning that they have the ability to move under their own power. At the beginning of the composting process (0-40°C), mesophilic bacteria predominate. Most of these are forms that can also be found in topsoil. As the compost heats up above 40°C, thermophilic bacteria takeover. The microbial populations during this phase are dominated by members of the genus *Bacillus*. The diversity of bacilli species is fairly high at temperatures from 50-55°C but decreases dramatically at 60°C or above. When conditions become un-favourable, bacilli survive by forming endospores, thick-walled spores that are highly resistant to heat, cold, dryness, or lack of food. They are ubiquitous in nature and become active whenever environmental conditions are favourable. At the highest compost temperatures, bacteria of the genus *Thermus* have been isolated. Composters sometimes wonder how micro-organisms evolved in nature that can withstand the high temperatures found in active compost. *Thermus* bacteria were first found in hot springs in Yellowstone National Park and may have evolved there. Other places where thermophilic conditions exist in nature include deep sea thermal vents, manure droppings, and accumulations of decomposing vegetation that have the right conditions to heat up just as they would in a compost pile. Once the compost cools down, mesophilic bacteria again predominate. The numbers and types of mesophilic microbes that recolonize compost as it matures depend on what spores and organisms are present in the compost as well as in the immediate environment. In general, the longer the curing or maturation phase, the more diverse the microbial community it supports.

Fungi

Fungi include molds and yeasts, and collectively they are responsible for the decomposition of many complex plant polymers in soil and compost. In compost, fungi are important because they break down tough debris, enabling bacteria to continue the decomposition process once most of the cellulose has been exhausted. They spread and grow vigorously by producing many cells and filaments, and they can attack organic residues that are too dry, acidic, or low in nitrogen for bacterial decomposition. Most fungi are classified as saprophytes because they live on dead or dying material and obtain energy by breaking down organic matter in dead plants and animals. Fungal species are numerous during both mesophilic and thermophilic phases of composting. Most fungi live in the outer layer of compost when temperatures are high. Compost molds are strict aerobes that grow both as unseen filaments and as gray or white fuzzy colonies on the compost surface.

Protozoa

Protozoa are one-celled microscopic animals. They are found in water droplets in compost but play a relatively minor role in decomposition. Protozoa obtain their food from organic matter in the same way as bacteria do but also act as secondary consumers ingesting bacteria and fungi.

Actinomycetes

The characteristic earthy smell of soil is caused by actinomycetes, organisms that resemble fungi but actually are filamentous bacteria. Like other bacteria, they lack nuclei, but they grow multicellular filaments like fungi. In composting they play an important role in degrading complex organics such as cellulose, lignin, chitin, and proteins. Their enzymes enable them to chemically break down tough debris such as woody stems, bark, or newspaper. Some species appear during the thermophilic phase, and others become important during the cooler curing phase, when only the most resistant compounds remain in the last stages of the formation of humus.

Actinomycetes form long, thread-like branched filaments that look like gray spider webs stretching through compost. These filaments are most commonly seen toward the end of the composting process, in the outer 10 to 15 centimetres of the pile. Sometimes they appear as circular colonies that gradually expand in diameter.

Composting

Compost is the product resulting from the controlled biological decomposition of organic material. Composting is the microbial decomposition of bio-degradable solid waste, under aerobic condition, where micro-organisms convert waste into a stable end product (compost). The term co-composting is also used to describe the composting process of two or more substances together. The main objectives of composting are to decompose organic fraction of waste to reduce its volume, weight and moisture content; minimize potential odor; decrease pathogens; and, increase potential nutrients for agricultural application. The composting process may minimize the spread of diseases because of the destruction of some pathogens and parasites at elevated temperature (Table 1).

TABLE 1
Destruction of some Common Pathogens and Parasite at Elevated Temperatures

Organisms	*Observations*
Salmonella typhosa	No growth beyond 46°C; death in 30 mm. at 55-66°C and 20 mm, at 60°C; destroyed in a short time in compost environment
Salmonella sp.	Destruction within 1 h at 55°C and 15-20 mm. at 60°C
Shigella sp.	Destruction within 1 h at 55°C
Escherichia coli	Destruction within 1 h at 55°C and 15-20 mm. at 60°C
Entamoeba histolytica cysts	Destruction within a few minutes at 45°C and within a few seconds at 55°C
Taenia saginata	Destruction within a few at 55°C
Necator americanus	Destruction within 50 at 45°C

Composting is similar to land treatment, but it can be more efficient. Also, with composting systems, treated waste is contained within the composting facility where its properties can be readily monitored. With composting, mixtures of the waste, soil (to provide indigenous bacteria), and other additives may be placed in piles to be tilled for aeration, or placed in containers or on platforms to allow air to be forced through the composting mixture. To optimize moisture conditions for biodegradation, the compost mixture is maintained at 40 to 60% water by weight. Elevated temperatures (30 to 70 degrees C) in compost mixtures increase microbial metabolism. (Mishra *et al.,* 1982).

The Composting Process

Composting is an aerobic process in which micro-organisms convert a mixed organic substrate into carbon dioxide (CO_2), water, and minerals and stabilized organic matter. Control of environmental conditions during the process distinguishes composting from natural rotting or decomposition (Zucconi and De Bertoldi, 1987). Controlled conditions, particularly of moisture and aeration are required to yield temperatures (120 to 140°F) conducive to the micro-organisms involved in the composting process (Chen and Inbar, 1993). The extent of organic matter decomposition at any particular time is related to the temperature at which composting takes place and to the chemical composition of the organic substrate undergoing composting (Levi-Minzi *et al.,* 1990). Because of the presence of readily degradable carbon (C), most organic materials initially decompose rapidly. Thereafter, decomposition slows because of the greater resistance to decomposition of remaining C compounds (lignin and cellulose), other environmental factors remaining constant. Generally, the higher the lignin and polyphenolic content of organic materials, the slower their decomposition (Palm and Sanchez, 1991).

Readily available (labile) organic nitrogen (N) is mineralized (converted to nitrate- N, a form that plants use) by microbial activity during the first weeks of decomposition. As the more labile organic N disappears, the most recalcitrant (resistant to microbial degradation) organic N predominates in the organic N pool, and the mineralization rate slows (Iglesias-Jimenez and meter to measure changes in oxygen concentration of an aqueous compost suspension), and other spectroscopic methods (e.g. nuclear magnetic resonance, gel chromatography, etc.).

Generally, proper conditions for active composting include an adequate supply of oxygen for microbial respiration (approximately 5% of the pore space in the starting material should contain air), a moisture content between 40 and 65%, particle sizes of approximately 1/8 to 2 inches in diameter, and a C : N ratio between 20:1 and 40 : 1. This first stage of composting lasts 1 to 2 days, during which time mesophilic strains of micro-organisms (species that are most active at temperatures of 90 to 110°F) initiate decomposition of readily degradable compounds. Sugars, fats, starches, and proteins are rapidly consumed, heat is given off and the temperature of the substrate rises. The pH typically decreases as organic acids are produced (Chen and Inbar, 1993).

The second stage is the thermophilic stage. When active composting is taking place, microbial activity in the pile should cause an increase in temperature in the center of the pile to about 120 to 140°F. When temperatures are within this range, specific, heat-loving (thermophilic) bacteria vigorously degrade organic material. Temperatures will remain in this range as long as decomposable materials are available and oxygen is adequate for microbial activity (Chen and Inbar, 1993).

Many important processes take place during the thermophilic stage. Organic matter is degraded and particle size is reduced. Pathogens are destroyed (above a critical temperature of 131°F). Fly larvae are killed and most weed seeds are destroyed at temperatures above 145°F. The pH frequently rises above as ammonia is liberated during protein degradation.

Temperatures normally remain in the thermophilic range for several weeks. Falling temperature is an indication that oxygen has become limiting for microbial activity and that the compost pile needs to be "turned" to reintroduce oxygen for renewed microbial activity. The pile can be re-mixed by hand, with a front-end loader, or with other specialized equipment to reintroduce oxygen to the pile or windrow. Alternatively, perforated pipe placed under the pile during construction can deliver oxygen through forced aeration from blowers and fans. Turning the pile also insures that materials are moved from outer portions of the pile where temperatures may be lower (than 1200°F) to inner portions where they will be subject to thermophilic temperatures. Several turnings usually ensure destruction of most pathogens, weed seeds, and insect larvae.

Temperature in the pile can fall for other reasons. For example, if moisture content falls below 40%, the pile may become too dry for microbial activity. With normal temperature in the pile as high as 140°F, evaporation of water is normal and re-wetting of the pile is often required. Temperature may also fall if the pile becomes too wet. If moisture content exceeds 65 to 70 %, much of the pore space in the pile will contain water rather than oxygen. Oxygen will quickly become limiting and microbial activity will decrease (reflected by decreasing temperature). Without sufficient oxygen the pile will become anaerobic, and an entirely different set of micro-organisms that function effectively without oxygen (anaerobes) will assume primary responsibility for decomposition. Unfortunately, this much slower process has many undesirable by-products, among them noxious odors [e.g. the "rotten egg" smell of hydrogen sulfide gas (H_2S)] and plant growth inhibitors (e.g. organic acids). A general rule of thumb is that the pile is too wet if water can be squeezed out of a handful of compost and too dry if the handful does not feel moist to the touch (Bull, 1999).

It is also possible for temperatures in the pile to become too hot. When temperatures reach the 150°F to 160°F range, thermophilic organisms begin to die and composting slows. If the pile is very dry, there is some danger of spontaneous combustion. After the pile has been turned several times, temperatures gradually fall to about 100°F. Active composting is completed, and the volume of the original material is normally reduced by 25 to 50%. Decomposition continues beyond this point but at a much slower rate, and little heat is generated. When the compost pile temperature falls to that of ambient air, the compost is ready for curing.

A normal curing period lasts for 30 days and helps insure against any negative consequences of application of immature compost in cropping situations; e.g. inhibition of seed germination. Less heat is generated during this period and the final pH is normally slightly alkaline. Common micro-organisms (pathogens and beneficial) as well as a micro fauna recolonize the compost. Intense microbial competition for food takes place through both direct antagonism and production of antibiotics (Chen and Inbar, 1993). It is in this competition that pathogens (e.g. *Pythium, Rhizoctonia, and Phytophthora* species) are often suppressed by beneficial microbial species (Hoitink and Fahy, 1986).

Composting is the biological decomposition of organic matter under controlled conditions brought about by the growth of micro-organisms and invertebrates. While decomposition occurs naturally, it can be accelerated and improved by human intervention. Composting stabilizes organic matter, yielding an end product that contains humus, and has a uniform crumbly texture. An understanding of the composting process is important for producing a high-quality product and preventing operating problems. Figure 1 illustrates the basic composting process. The micro-organisms and invertebrates that decompose manure and other farm wastes require oxygen and water, and produce compost, carbon dioxide, heat and water (Martin and Gershung, 1992). The organic wastes provide nutrients (nitrogen and carbon) necessary for micro-organisms to carry out decomposition efficiently.

The heat produced increases the temperature in the compost pile from near-ambient air temperature to as high as 70°C (160°F). The temperature rise results in increased water evaporation. As the process nears completion (after one month to one year), the compost pile once again approaches ambient air temperature. Composting leads to a volume reduction. Much of this reduction results from the loss of carbon dioxide, water and other minor gases to the atmosphere.

Decomposers

Naturally occurring micro-organisms and invertebrates are the primary decomposers that accomplish composting. These micro-organisms include bacteria, mould or fungi, actinomycetes and protozoa. (Chang and Hudson, 1967) Tiny invertebrate animals such as mites, millipedes, insects, sow bugs, earthworms and snails are the primary agents of physical decay.

They break up waste debris and transport micro-organisms from one site to another. The ease with which organic materials are composted depends on the type of decomposers, the type of organic material being composted and the composting method used. For example, many decomposition organisms can utilize the carbon in sugar found in straw, while fewer decomposers can use the carbon in cellulose or lignin fibers found in paper or wood. As carbon compounds decompose decomposers can use the carbon in cellulose or lignin fibers found in paper or wood.

Organic Residues

Leaves, Grass clippings, Other plant debris, Food scraps, Fecal matter and animal bodies including those of soil invertebrates

↓

Primary Consumers

(Organisms that eat organic residues)

Bacteria, Fungi, Actinomycetes, Nematodes, Some types of mites, Snails, Slugs, Earthworms, Millipedes

↓

Secondary Consumers

(Organisms that eat primary consumers)

↓

Tertiary consumers

(Organisms that eat secondary consumers)

Centipedes, Predatory mites, Beetles, Ants.

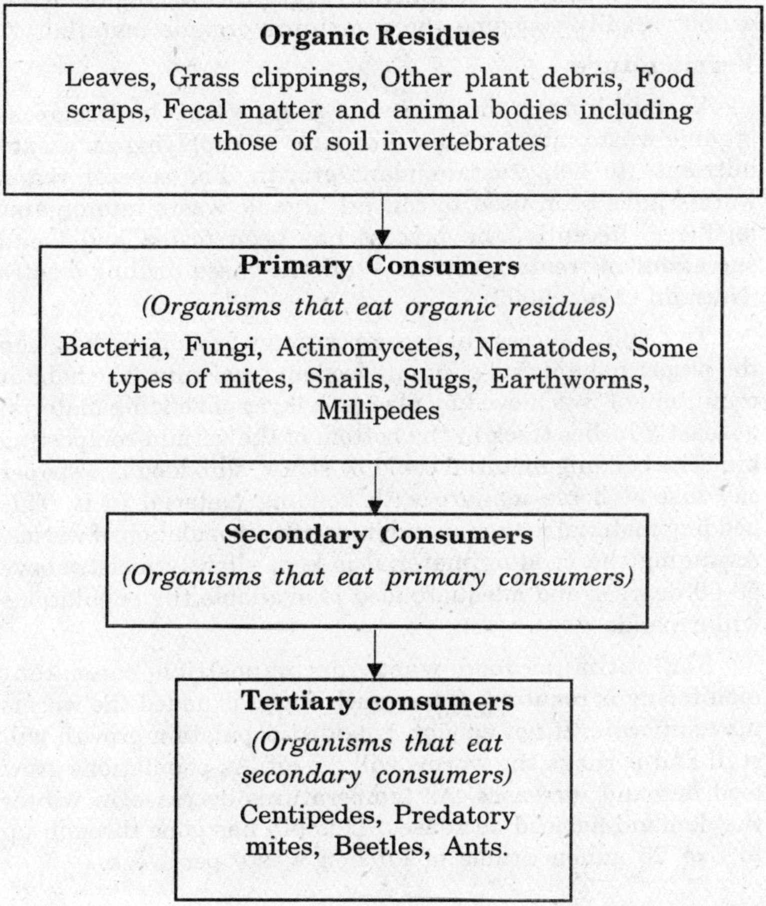

As carbon compounds decompose, part of the carbon is converted to microbial and invertebrate cell structure, while most is converted to carbon dioxide, which is lost to the atmosphere. Different decomposers prefer different organic materials and temperatures; therefore, the more diverse the microbial populations, the better. If the environment becomes unsuitable for a decomposer, that organism will become dormant, die or move to a more hospitable area of the pile. Changing conditions during the compo sting process lead to an ever-changing ecosystem of decomposition organisms. Decomposer activity diminishes when the micro-organisms cannot readily consume the remaining organic material.

Vermiculture

Vermiculture is the process of using worms to decompose organic waste into a material capable of supplying necessary nutrients to help sustain plant growth. For several years, worms have been used to convert organic waste into organic fertilizer. Recently, the process has been tested and found successful in treating certain synthetic-based drilling wastes (Norman *et al.*, 2002).

To achieve successful decomposition worms must be given the proper habitat and adequate food and moisture. The habitat requirement is achieved by placing a layer of bedding material at least 3 inches thick in the bottom of the vermin-composting bin. The bedding material could be straw, shredded newspaper sawdust or horse manure with bedding material in it. This bedding material is then inoculated with a population of worms. Assuming the bedding material is kept slightly moist, above 50-60 degrees and adequate food is available the populations will expand.

For optimum food waste decomposition consistent monitoring is required. If too much waste is added the worms may suffocate. If not enough is added population growth will stall and perhaps the worms will die-off. As populations grow food demand increases. As temperatures decrease in winter the demand for food decreases. This bin has gone through up to five 25 gallon drums of kitchen waste per week.

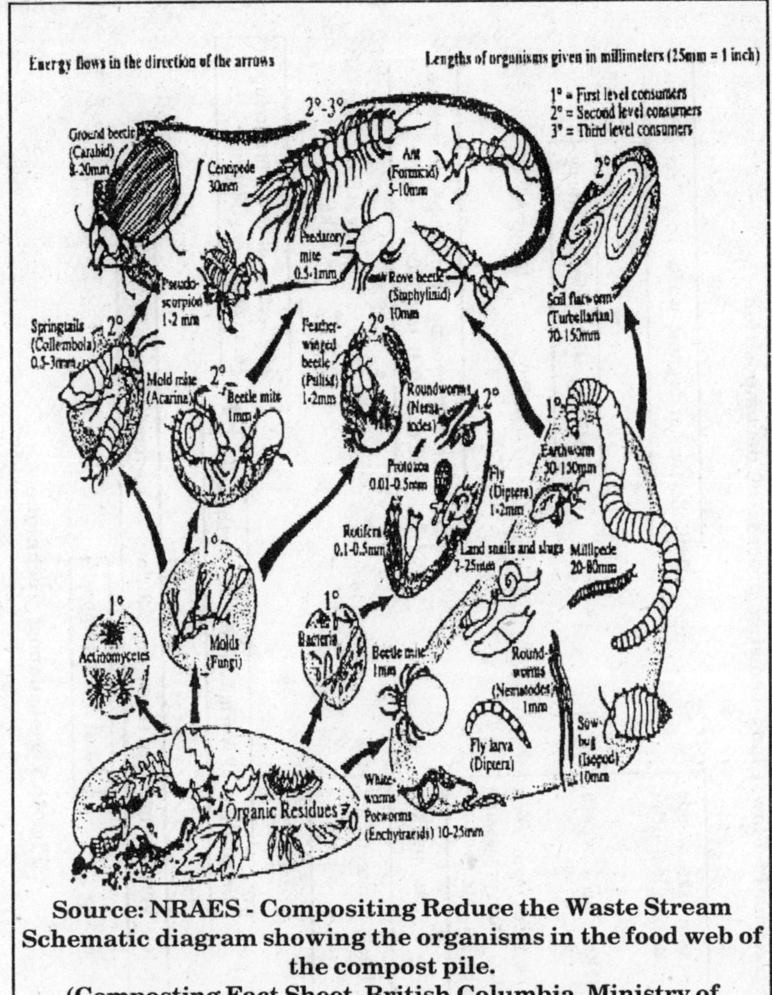

Source: NRAES - Compositing Reduce the Waste Stream
Schematic diagram showing the organisms in the food web of the compost pile.
(Composting Fact Sheet, British Columbia, Ministry of Agriculture and Food September, 1996).

Factors Effecting Microbial Decomposition

Moisture

Water is necessary for the microbes to work efficiently, as it acts as a medium for chemical reactions, the transport

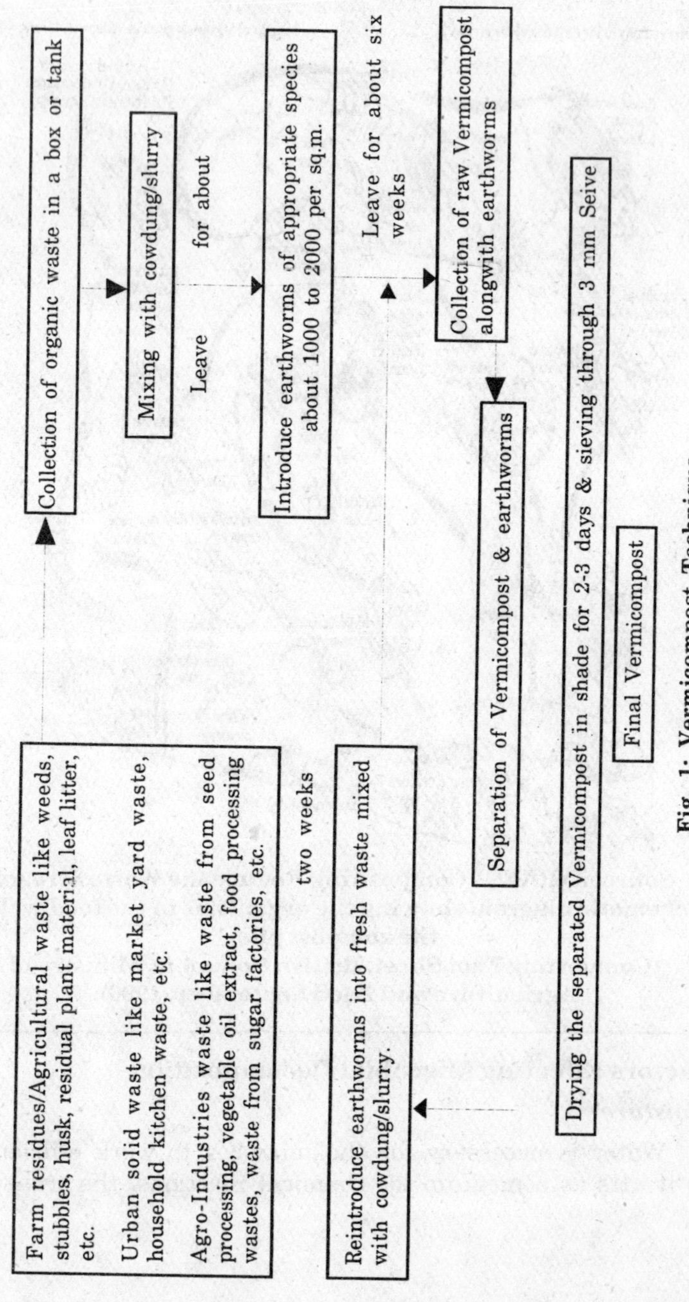

Fig. 1: Vermicompost Technique

of nutrients and the movement of the micro-organisms. The moisture content of the compo sting material should be maintained between 40% and 50% (wet basis). Too much moisture will cause conditions to become anaerobic (lack of oxygen) and unpleasant odour may result. The compost is too wet if water can be squeezed out by hand, or too dry if it is not moist to the touch. The composting material generally dries out with time, thus water should be added regularly. To ensure the water is well distributed within the material this is ideally done when turning.

The moisture content of the compost varies depending on the porosity of the reactor feed, free air space, aeration, temperature and other related physical factors. Moisture in this context is defined as weight loss after the sample has been dried to constant weight at 11-110°C. Water activity which specifically addressed the available water for microbial activity and for chemical reaction, is not commonly used to describe water relationships in composting. The optimum moisture in composting also depends on the strength of the reactor feed to withstand compaction which changes physical characteristics of compost layers, ultimately resulting in decreased aeration and microbial activity.

Excessive moisture inhibits aerobic metabolism as a result of oxygen diffusion limitation, a lack of water also impedes composting. The optimal moisture content in composting has been empirically determined to range between 50 and 60% for bacteria (Golueke, 1972 and Poincelot ,1972). Bacterial metabolic activity is severely limited when moisture content drops below 40%. Fungi have a lower moisture threshold. Snell (1957) showed that oxygen uptake during composting at moisture levels below 30% was approximately 15% of that at 55% moisture. Miller *et al.* (1985) used water potential to characterize the composting process. He stated that at water potentials below -20kpa (about 60% moisture), bacteria progressively failed to colonize the compost mass.

Carbon to nitrogen ratio (C:N)

Micro-organisms require a balance of carbon and nitrogen for healthy cell growth. It is important to provide carbon and nitrogen in the right proportions to encourage microbial activity.

A carbon to nitrogen ratio (C: N) between 15: 1 (15 parts carbon to 1 part nitrogen) and 30:1 is required for good composting results. While the C: N largely determines the blend of materials to be used, the rate at which carbon compounds decompose should also be considered.

The optimal carbon /nitrogen ratios for the microbiological decomposition of organic material in composting process have been reported to be in the range of 26 to 35 (Poincelot, 1972). In general, this range is similar to that reported for agricultural soils. The underlying biological basis of the optimum ratio is obscure. Microbial biomass, on a dry weight basis, contains approximately 50% carbon and 10 to 14% nitrogen. Allowing for respiration (i.e. carbon substrate as the electron donor), the optimal ratio may reach values as high as 30 as reported for composting. Grass clipping have an elemental carbon/nitrogen ratio of about 20, and leaves have a ratio between 40 and 80 (Poincelot, 1972). If the carbon/nitrogen ratio is low, as is the case for grass clippings, the microbiological degradation leads to excess ammonia formation, which increases the pH and thereby enhances ammonia volatilisation. Conversely, if the carbon/nitrogen ratio is too high , the process becomes nitrogen limited, nitrogen limitation and its effect need to be better documented for composting systems.

Aeration (Oxygen Addition)

The composting pile should contain sufficient oxygen to maintain adequate microbial activity. While the initial mixing of the materials will introduce oxygen into the pile, this small supply of oxygen will be rapidly exhausted. To maintain sufficient oxygen levels, some form of aeration will be required. Turning the piles regularly (weekly) is recommended for good aeration, as the pores created allow air to move easily through the pile. Air could also be introduced to the material via forced aeration. A bulking agent such as sawdust or woodchips can be used to increase the air pockets within the composting material.

Decomposition is concerned primarily with the biological oxidation of organic waste material of recent origin via microbial metabolism to a stabilized organic residue. The process is associated with the production of heat, microbial biomass, CO_2 and H_2O. it is desired that the composting process be based on aerobic decomposition and thus the availability of oxygen to the compost process is prime importance.

Carbon and Nitrogen are the two most important elements required for microbial decomposition:

Carbon:	Energy source and the basic building block of microbial cells (more than 50 per cent).
Nitrogen:	Crucial component of the proteins, nucleic acids, amino acids, enzymes and co- enzymes necessary for microbial cell growth and function
Ideal C/N ratio for composting:	30 parts carbon for each part nitrogen by weight (C:N ratio 30:1)
C:N<30:1	Nitrogen will be supplied in excess and will be lost as ammonia gas, causing undesirable odors
C:N>30:1	Not sufficient nitrogen for optimal growth of the microbial populations -> compost will remain relatively cool and degradation will proceed at a slow rate.

The microbiological decomposition of pesticides in top-soils occurs largely via aerobic metabolism. For aromatic pesticides, ring activation and fission involve oxygenase activities to produce central metabolites. Mechanisms for anaerobic degradation of toxic organic compounds involve reductive dehalogenation before ring activation (Rochkind *et al*., 1987, Tiedje, 1982 and Wallnofer Engelhardt, 1989).

Jeris and Regan (1973) suggested that 30-36% free air space is required to obtain adequate aeration for composting for a wide variety of materials. Free space is a derived parameter, calculated on the basis of moisture. A study has yet to be reported to determine the rate of composting in relation to variation in the free air space of the compost. There have been some attempts to develop a mathematical relationship between oxygen consumption rates and the temperature for composting process.

Particle size

A small particle size will ensure the microbes have ready access to the organic material to help speed up the composting process. Smaller particles also allow oxygen to penetrate through the pile more evenly and be available to the microbes. If the material is clumped into large particles, the material will be compacted, allowing less oxygen into the pore spaces.

Temperature

The natural composting processes produce heat, and the micro-organisms grow best within a temperature range between 55°C and 65°C. Operation at this temperature also assists the destruction of pathogens and weed seeds. However if the temperature exceeds 65°C the composting organisms will die and the process will slow down substantially. Provided the oxygen and water supplies are within the optimum ranges, the composting material should maintain suitable operating temperatures. In Queensland, composting can be successfully achieved throughout the year. However due to the higher temperatures during summer, more moisture addition may be necessary in drier areas.

pH

A pH value between 5.5 and 8.5 is optimal for compost in micro-organisms as bacteria and fungi digest organic matter, they release organic acids. Optimal pH values for composting range from pH 5.5 to 8.5. Bacteria favour a near neutral pH, where as fungi favour an acidic pH range. The pH of the yard waste composted in the Portland study ranged between pH 5.8 and 7.2 (Gurkewitz, 1989).

The effect of extreme pH on the composting process are directly related to pH on the microbial activity or more specifically, on microbial enzymes. The pH changes in the yard waste are sequential initially the pH drops further as a result of acid forming bacteria it then increases and becomes alkaline and finally drops back to near neutral as a result of humus formation the pH buffering capacity increases as a result of humus formation (Poincelot, 1972).

The Use of Effective Microorganisms (EM) in Organic Waste Management

EM owes its discovery to the work of Dr. Terno Higa, a microbiologist and organic farmer from the University of the Ryukyus in Okinawa, Japan, who made an accidental discovery while researching the various beneficial aspects of isolated strains of micro-organisms on soil composition and plant growth. In 1982, Dr. Higa learned that when blended in a state of balance, certain mixtures of beneficial micro-organisms promoted "healthy plant growth resulting in more abundant harvests of better tasting crops" (Higa,1993). After months of testing these mixtures on seasonal crops of mandarin oranges with entirely positive results, Dr. Higa was sure that he had made an astonishing discovery.

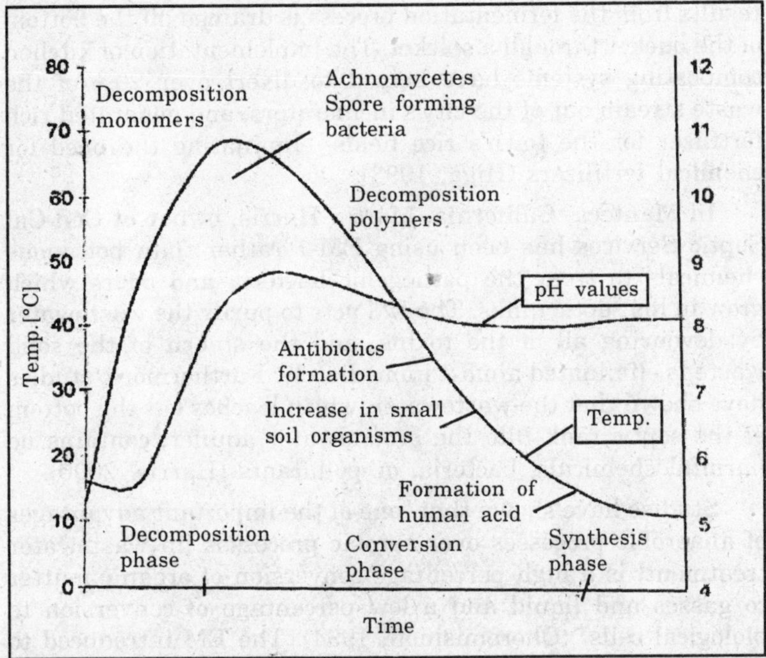

New Horizons for Municipal Waste Management

Kani City, Japan has become a model city for the effective widespread use of EM on a municipal level. In 1992, the suburban city was at its wits end over trash disposal issues, and turned to the EM Research Organization for help. Incinerators were running at full capacity, spewing toxic fumes across the entire town. To begin the EM treatment process, all households were given EM home composting buckets to use in their kitchens. People simply mix five gallons of food waste with EM- bokashi, seal the lid to ensure an anaerobic environment, and in four weeks time, the food has decomposed into an odorless fermented fertilizer. The EM-rich liquid that results from the fermentation process is drained off the bottom of the bucket through a spicket. The implementation of kitchen composting systems have helped to divert over 20% of the waste stream out of the city's incinerators, and generated rich fertilizer for the town's rice fields, eliminating the need for chemical fertilizers (Higa, 1993).

In Manteca, California, Martin Harris, owner of Cen-Cal Septic Services has been using EM-1 rather than poisonous chemicals to treat the pathogenic bacteria and odors which grow in his septic tanks. The EM acts to purify the wastewater by devouring all of the toxins, and the stench of the solid waste is eliminated almost immediately. Furthermore, studies have shown that the wastewater, which leaches out the bottom of the septic tank into the groundwater aquifer, contains no harmful chemicals, bacteria, or pollutants (Harris, 2000).

Studies have shown that "one of the important advantages of anaerobic processes over aerobic processes [in wastewater treatment] is a high percentage conversion of organic matter to gasses and liquid and a low percentage of conversion to biological cells" (Cheremisinoff, 1994). The EM introduced to anaerobic wastewater treatment facilities help to reduce the unpleasant by products of anaerobic decomposition, leaving very little residual sludge (Cheremisinoff, 1994).

Recently, the EM Research Organization (EMRO) of Tucson, Arizona has introduced the use of EM to large scale agricultural and livestock operations across the USA. Examples include the deodorization of Tyson Chicken's vast poultry and pork farms in Missouri. Not only has the EM helped to devour the malodors associated with intense livestock operations, in has also been used to compost vast amounts of animal waste into high-grade fertilizer sold to local markets (Wood, 2000). Scientific purists is the fact that the introduction of EM into soils alters the natural chemical composition of the soils, displacing native micro-organisms and nutrients, which may be harmful to the survival of native plant species (Smith, 1992).

Bioreactors

Bioreactors work according to the same aerobic biological reactions that occur in land treatment and composting, but the reactions occur in an open or closed vessel or impoundment. This environment accelerates the rate of biodegradation by allowing better control of the temperature and other conditions that affect the biodegradation rate. Bioreactor processes are typically operated as a batch or semi-continuous process. In a bioreactor, nutrients are added to slurry of water and waste, and air sparging or intensive mechanical mixing of the reactor contents provides oxygen. This mechanical mixing results in significant contact between micro-organisms and the waste components being degraded. To accelerate system start-up, introduction of microbes capable of degrading the organic constituents of the waste may be useful, although some companies have not had favorable experience with designer bugs. Many of the additives used for bioreactors are common agricultural products and plant or animal wastes.

In tank-based bioreactors, operating conditions (temperature, nutrient concentration, pH, oxygen transport and mixing) can be monitored and controlled easily. Optimized biological processes ensure the best rate of biodegradation and allow for reduced space requirements relative to land-based biological treatment processes. However, capital and operation and maintenance costs for bioreactors are high relative to other forms of biological treatment. McMillen *et al.* (2004) reported estimated costs for bioreactor treatment of oily cuttings wastes of approximately $500 per cubic metre.

REFERENCES

Adams, M.W.W. and R.M. Kelly. Enzymes Isolated from Microorganisms that Grow in Extreme Environments. *Chemical and Engineering News* 73, 32-42, 1995.

Bull, E.L., The value of dead wood to vertebrates in the Pacific Northwest. *In: The Ecology and Management of Dead Wood in Western Forests*. November 2-3 1999. Reno, NV, 1999.

Chang, Y., Hudson, H.J., The fungi of wheat straw compost. In: *Ecology Studies. Trans. Br. Mycol. Soc*. 50: 649-66, 1967.

Cheremisinoff, Paul N., *Biomanagement of Wastewater and Wastes*. Englewood Cliffs, NJ.: Prentice-Hall, Inc., 1994.

Chen, Y., and U. Inbar, "Chemical and spectroscopical analyses of organic matter transformation during composting in relation to compost maturity". In: H.A.J. Hoitink and H.M. Keener (Eds.) *Science and Engineering of Composting: Design, Environmental, Microbiological and Utilization Aspects*. pp. 550-600, 1993.

Ference, Don *et al.*, Market Study and Strategic Plan for Composted Hog Manure, *Report for Canada-British Columbia Subsidiary Agreement on Agri-Food Regional Development*, Victoria, 1989.

Golueke, C.G., *Composting: A Study of the Process and its Principles*, Rodale Press, Inc., Emmaus, PA., 1972.

Gurkewitz, S., Yard waste compost testing. *Biocycle* 30(6): 58-59, 1989.

Harris, Martin, Personal interview. May 1, Owner, Cen-Cal Septic Services, Manteca, California, 2000.

Higa, Teruo, *An Earth Saving Revolution: A Means to Resolve Our World's Problems through Effective Microorganisms* (EM). Tokyo: Sunmark Publishing, Inc., 1993.

Hoitink, H.A.J. and P.C. Fahy, Basis for the control of soilborne plant pathogens with composts. *Annual Review of Phytopathology*. 24:93-114, 1986.

Hoitink, H.A.J. and M.J. Boehm, Biocontrol within the context of soil microbial communities: A substrate-dependent phenomenon. *Annual Review of Phytopathology*. 37: 427-446, 1999.

Jeris, J.S. and W.R. Regan, Controlling environmental parameters of optimum composting. Moisture, free air space and recycle. *Compost Sci.* 14(2): 8-15, 1973.

Kimmins, J.P. Organic Wastes as Forest Fertilizers, *BioCycle*, Emmaus, PA, 1991.

Levi-Mintz, R., R. Riffaldi, and A. Saviozzi. Carbon mineralization in soil amended with different organic materials. *Agriculture, Ecosystems and Environment.* 31: 325-335, 1990.

Maser, C., R.G. Anderson and K. Cromack, Jr., Dead and down woody material. Pages 78-95 In: *Wildlife habitat in managed forests: The Blue Mountains of Oregon and Washington. USDA For. Serv., Agric. Handbook.* 253, 1979.

Mishra, M.M., Kapoor, K.K., Yadav, K.S., Effects of compost enriched with Mussoorie rock phosphate on crop yield. *Indian J. Agric. Sci.* 52: 674-678, 1982.

Martin, Deborah L. and Grace Gershuny (eds.), *The Rodale Book of Composting*, Rodale Press, 1992.

McMillen, S.J., R. Smart, R. Bernier, and R.E. Hoffman, "Biotreating E&P Wastes: Lessons Learned from 1992-2003," SPE 86794, presented at the *Seventh International Conference on Health, Safety, and Environment in Oil and Gas Exploration and Production*, Calgary, Alberta, Canada, March 29-31, 2004.

Miller, R.L., J.L.W. Keularts, J.R. Street, W.E. Pound and W. Shane, *Control of Turf Grass Pests. Ohio Cooperative Extension Service Leaflet 187.* The Ohio State University, Columbus, 1985.

Norman, M., S. Ross, G. McEwen, and J. Getliff, "Minimizing Environmental Impacts and Maximizing Hole Stability - the Significance of Drilling with Synthetic Fluids in New Zealand," *New Zealand Petroleum Conference Proceedings*, February 24-27, 2002.

Palm, C.A., and P.A. Sanchez, Nitrogen release from the leaves of some tropical legumes as affected by their lignin and polyphenolic contents. *Soil Biology and Biochemistry.* 223: 83-88, 1991

Poincelot, R.P., The biochemistry and methodology of composting. *Connecticut Agricultural experimental Station Bulletin 127.* Connecticut Agricultural Experimental Station, New Haven, 1972.

Rochkind-Dubinsky, S.L., G.S. Sayler, and J.W. Blackburn, *Microbiological Decomposition of Halogenated Aromatic Compounds.* Marcel Dekker Inc., New York, 1987.

Smith, S.R. Sewage sludge and refuse composts as peat alternatives for conditioning impoverished soils: Effects on the growth response and mineral status *of Petunia grandiflora. Journal of Horticultural Science.* 67: 703-716, 1992.

Snell, J.R. Some engineering aspects of high rate composting. *J. Sanit. Eng. Div. Proc. Am. Soc. Civ. Eng.* 83: 1178-1-1178-36, 1957.

Tiedje, J.M. Identification of thermophilic bacteria in solid waste composting. *Appl. Environ. Microbial.* 50: 906-915, 1982.

Wood, Matthew. Personal Interview. April 1, Member. EM Research Organization, Okinawa, Japan. Owner, Sustainable Community Development, Columbia, Missouri, 2000.

Wallnofer, P.R., and G.Engelhardt. "Microbial degradation of pesticides". In: G. Haug and H. Hoffmann (eds.), *Chemistry of Plant Protection*, Vol.2. Springer-Verlag KG, Berlin, 1989, pp. 1-115.

Zucconi, F. and M. de Bertoldi, "Compost specifications for the production and characterization of compost from municipal solid waste". In: M. de Bertoldi *et al.* (Eds.), *Compost: Production, Quality and Use*, Elsevier Applied Science: London, 1987, pp. 30-50.

CHAPTER 8

Integrated Pest Management in Horticultural Crops: A Strategy for Sustainable Environmental Protection

S.C. Swain* and L.R. Patra**

ABSTRACT

Horticultural crops receive heavy insecticides inputs for higher production and productivity. This has resulted in widespread residues and health hazard, insecticide resistance, resurgence of pest problems, loss of bio-diversity and overall disruption of natural agro-ecosystem. Efforts have been made to standardize low-cost, eco-safe and sustainable Integrated Pest Management strategies against many economical crops. However, these are still insecticide based blended with compatible combination of cultural and biological strategies. But this is no way reflects non-feasibility of non-insecticide IPM modules. Now, 100% effective insecticide free IPM packages for most notorious pests of horticultural crops like cabbage, tomato, cucurbits, mango and many other crops have been evolved. As the pesticide residues is likely to become a serious non-tariff barrier, no insecticide based IPM packages need to be widely adopted in 143 million hectares cultivable area of India for export of more farm produce.

* Scientist (Horticulture) Krishi Vigyan Kendra, Rayagada, Orissa University of Agriculture & Technology, At/Po. Gunpur, Distt. Rayagada, Orissa-765022.

** Lecturer (Zoology), Laboratory of Environmental Toxicology, Dept. of Zoology, Aska Science College, Aska, Ganjam, Orissa.

Introduction

Several thousands of insect pests have been recorded to feed on various fruit, vegetable, spice and plantation crops in India. More than 200 insect species are reported on mango alone in this country (Shukla *et al.*, 2001). However, the number of key pests causing serious yield loss on different crops under an agroclimatic region is restricted only to few species. Pockets of intensive and commercial crops receive heavy insecticide input due to more quality considerations. Vegetables alone are targeted to 15 per cent of the national insecticide consumption. This has resulted in widespread residues and health hazard, insecticide resistance, resurgence of pest problems, loss of biodiversity and overall disruption of natural agro-ecosystem.

Efforts have been made in past 25 years to synthesize low-cost, eco-safe and efficient integrated pest management (IPM) strategies against many economically important pests and crops. The approximate number of pests, key problems and the nature of IPM developed for different crops have been presented in Table 1. An analysis of situation reveals that most of the management packages recommended against pests/crops are still insecticide based blended with compatible combination of cultural and biological strategies. But this in no way reflects non-feasibility of non-insecticide IPM modules. We have now effective 100 per cent insecticide-free packages for most notorious pests of horticultural crops like cabbage, tomato, cucurbits, mango and number of other crops. Some of these modules against problem pests are given below.

Some Examples of IPM Modules

IPM for Insect Pests in Cabbage (Krishna Moorthy and Sardana, 2001)

Problems and Hot Spots

1. Diamond back moth *(Plutella xylostella)*: Worldwide
2. Aphid *(Brevicoryne brassicae)*: All over India.
3. Leaf webber *(Crocidolomia binotalis)*: Karnataka
4. Cabbage stem borer *(Hellula undalis)*: Karnataka
5. Tobacco caterpillar *(Spodoptera litura)*: U.P.

IPM Module

1. Grow 2 rows of bold seeded mustard as trap crop for DBM per 25 cabbage rows. So first row of mustard 15 days before cabbage transplanting and the second, 25 days after transplanting.
2. Spray 4 per cent NSKE at head initiation i.e. between 17-28 DAP depending on variety used if the population of DBM is @ 0.5/Plant.
3. Apply 2 to 3 additional sprays with 4 per cent NSKE at 10-15 days interval. 2 to 3 repeated applications of Bt. @ 5000 g/ha. at 15 days interval can alternatively be used for NSKE.
4. Apply dichlorvos 0.1 per cent on mustard 9-15 days after sowing to kill trapped insects.

Thus, cabbage pests can be managed exclusively by NSKE. However, its treatments should not exceed 4 sprays and the application after the formation of curd should be limited to maximum of one spray so as to prevent adverse effect on quality. Similarly, its treatment on very young plant should not be done for phytotoxicity reasons. Efficiency of home made preparation is higher as compared to many commercial products.

Integrated disease Management in Cabbage (NCIPM, 2001)

Key Diseases and Hot Spots

1. Downy mildew *(Peronospora parasitica)*: U.P.; Karnataka.
2. Bacterial black rot *(Xanthomonas comprestris pv. compestris,)*: Near big towns.
3. Reniform nematode *(Rotylenchulus reniformis)*: All cabbage growing areas.
4. Root knot nematode *(Meloidogyne incognita)*: All chickpea growing areas.

(a) Nursery

1. Soil solarization of bed with 60-100 gauge polythene sheet 2-3 week before sowing.
2. Addition of 50 g of *Trichoderma harzianum* (106 spores/ ml) to FYM to prevent soil and seed borne diseases.

TABLE 1
Important insect pest problems on vegetables, Fruits, Spices and Condiments and Plantation Crops

Crops	*No. of recorded pests*	*Key problems*	*Management module available*
A. Vegetables			
Potato	33	*Agrotis sp., Epilachna, aphid complex,* tubermoth.	Insecticide based
Brinjal	53	Shoot & fruit borer, coccinellid beetle, gall midge, red spider mite.	Insecticide based
Tomato	15	Fruit borer, whitefly, leaf miner, white grub	Insecticide-free
Chilly pepper, Bell pepper	—	Chilly thrips, mite spp., peach aphid, thrips, mite, gall midge.	Insecticidal
Okra	36	*Amrasca sp., Earias spp., Aphis gossypii*	Insecticidal
Cabbage	26	DBM, leaf webber, stem borer, aphid and *Spodoptera litura*	Insecticide-free
Cauliflower	19	-do-	Insecticide-free
Radish	11	Sawfly, aphids	Insecticide-free
Cucurbits	58	Raphidopalpa, Dacus cucurbitae	Insecticide-free
Sweet potato	20		
Drum stick	27	Blossom midge, bark caterpillar, flower	Insecticidal thrips
French bean	16	*Ophaseoli, Acyrthosiphum,* Aphis, gossypii	Insecticidal

Peas	18	Beanfly, pea aphid, leaf miner, pod borer.	Insecticidal
Country beans	38	Adisura	Insecticidal
B. Fruits			
Banana	41	Rhizome weevil, pseudqstem weevil, leaf caterpillar	
Ber	80	Fruit fly, fruit borer, mealy bug and scales, bark caterpillar	
Citrus	250	Psylla, leaf miner, whiteflies, lemon butterfly, sucking moths	
Custard apple	20	Mealy bug	
Fig	50	Fruit flies, gall maker, scales and mealy bugs	
Grape*	85	Shoot hole maker, leaf roller, mealy bugs and thrips	
Guava	80	Fruit fly, bark caterpillar, aphids, scales and mealy bugs	
Litchi	50	Fruit borer and mites	
Mango*	200	Leafhoppers, mealy bugs, shoot gall maker, stone weevil, leaf cutting weevil and shoot borer	
Papaya	15	Fruit flies, aphids and grasshopper	
Pine apple	>2	Mealy bugs	
Pomegranate	48	Fruit borer, aphids, mealy bugs and scales	
Sapota	33	Bud and fruit borer, Fruit flies,scales and mealy bugs.	
*Apple, Peach, Plum, Pear	217	Sanjose scale, woodly aphid, borer complex, fruit sucki moths, codling moth and peach curl aphid.	
Aonla	>12	Aphid, leaf twister, hairy caterpillars, shoot gall maker, Fruit borer and bark caterpillar.	

(contd.)

Crops	*No. of recorded pests*	*Key problems*	
C. Spices and Condiments			
Onion	9	Thrips	
Garlic	5	Thrips	
Large cardamom	3	Aphids and leaf caterpillar	
Chillies	14	White fly, thrips, fruit borers	
Coriander	4	White fly and aphids	
Turmeric	8	Thrips, stem borer	
Cardamom	31	Aphids, thrips, bud worm, looper caterpillar, beetles	
Pepper	13	Scales, thrips, shoot borer and beetles	
Ginger	4	Turmeric skipper, fly maggot.	
D. Planta tion Crops			
Arecanut	15	Arecanut mired bug	*Carvolhoia areceae* Miller and China
Cashewnut	47	Cashew tea borer	*Plocaederusferrugineus* Linn.
Coconut	41	Coconut scale Black	*Aspidiotus destructor* Signaret

		headed caterpillar	*Opisina arenosella* Walker
		Rhinoceros beetle*	*Oryctes rhinoceros* Linn.
		Red palm weevil	*Rhynchophorusferrugineus*
		Coconut weevil	*Olivier Diocalandrafrumenti* (F.)
		Coconut white-grub.	*Leucopholis coneophora* Burm
Coffee	48	Striped mealy bug	*Ferrisia virgata* (Cockerell)
		Soft green scale	*Coccus viridis* (Green)
		Nelmet scale	*Saissetia coffeae* (Walker)
		Coffee stem borer	*Xylotrechus quadripes* Chevrolat
		Coffee shot-holer borer	*Xylosandrus campactus* (EichhofJ)
		Coffee berry borer	*Hypothenemus hampei* (Ferrari)
Tea	156	Tea mosquito bug	*Helopeltis theivora* Waterhouse
		Bunch caterpillar	*H antonii Signoret* Andraca
		Humped slug	*Spatulicraspeda bipunctata*
		caterpillar	*Walker castaneiceps* Hamps
		Yellow tea mite	Polyphagotarsonemus latus (Banks)

Source: Tondon (1994); Nayar et al. (1976).

* *Insecticide-free IPM Package developed*

3. Soaking 1 kg seed/100 ppm streptocycline sulphate solution for 15 minutes before sowing to check black rot infection.
4. Using nylon nets in nursery beds to check entry of virus vectors.

(b) Main Field

1. At early stage at insect density of 0.5/plant, spraying of Bt @ 500g /ha twice at 15 days interval (additional 2 sprays of NSKE 0.5% may be done).
2. 4-5 release of *Trichogramma bactrae* @ 0.5-0.75 lakh adults/ha. at weekly interval, the first to start 20-30 DAT to control DBM.
3. Removal of Alternaria affected button leaves periodically and in case of heavy Alternaria attack, spraying of chlorothalonil 0.2 per cent.
4. Removal of black rot affected heads periodically and in case of severity, spraying of blitox 0.2 per cent +100 ppm streptocycline sulphate.
5. Destruction of crop residues after harvest.

This led to increase in parasite, *Cotesia plutellae* population to 4.31 pupae/plant as compared to 0.84 in non-IPM plots. Cost-benefit ratio was 1.1.6 in favour of IPM excluding the cost of production.

IPM for insect Pests in Tomato (NCIPM, 2001)

Problems and Hot spots

1. Fruit borer *(Helicoverpa armigera)*: All chickpea growing areas.
2. Leaf curl vectors: All chickpea growing areas.

IPM Module

1. Nylon net covered nursery raising to prevent entry of virus vectors.
2. Trap cropping through planting of 40 days old seedlings of tall marigold cv. Golden Age with yellow/orange flower colour, one row followed by 14 rows of 25 days old tomato seedlings. Marigold is effective to reduce

the incidence of serpentine leaf miner and results in negligible egg laying of fruit borer on tomato.

3. Application of NPV at the rate of 500 l.e. at flower initiation stage and to be repeated 2-3 times at weekly interval. NPV is mixed with 2 per cent jaggery and sprayed during evening hours.
4. Either of the above two measures (items 2 and 3) are effective to manage fruit borer population. Incase of nursery, 5-6 release of *Trichogramma brasiliensis* at the rate of 40,000 adults/ha. each time may be done.

Integrated Diseases Management Package in Tomato

Key Diseases and Hot Spots

1. Early blight *(Alternaria solani)* : All over tomato growing areas.
2. Damping off *('Pythium inflatum)* : All over tomatos growing areas.
3. Leaf curl *(Gemini virus)* : All over tomatos growing areas.
4. Bacterial leaf spot *(Xanthomonas compestris)* : All over tomatos growing areas.

(a) Nursery

1. Soil solarization of nursery bed and main crop field against damping off, collar rot, nematode etc.
2. Seed treatment with *Trichoderma harzianum* at the rate of 4 g/kg seed.
3. Seed treatment with imidacloprid at the rate of 3 g/ kg seed for protection from whitefly.
4. Nursery bed covering with 200-gauge nylon or muslin net to prevent entry of whitefly.
5. Soil drenching with captan at the rate of 0.3 per cent along with soil application of bioagents or neem cake.

(b) Main Field

1. Planting of one row of marigold cv. Golden Age for every 16 rows of tomato as trap crop against fruit borer's oviposition.

2. 2-3 spraying of NSKE (4%) at 10 days interval.
3. Erecting bird perches at the rate of 50/ha.
4. Releasing *Trichogramma pretiosum* at the rate of 50,000 adults/ha per release 5 times at weekly interval soon after noticing eggs of fruit borer which are seen mostly on minor 3 top leaves.
5. Spraying of NPV at the rate of 250 l.e./ha. (1.5 × 1012 POB) 2-3 times at weekly interval during evening.
6. Spraying twice streptocycline at the rate of 150 ppm followed by blitox-50 at the rate of 0.3 per cent to control Ahernaria, collar rot and bacterial diseases.

A Successful IPM Validation Package on Tomato at PDBC, Bangalore

1. Nursery sprayed with copper oxychloride (for Alternaria) and covered with nylon net (for whitefly).
2. Seedling root dip with imidacloprid (0.05%) fo'r whitefly and transplanting at 60 × 90 cm spacing keeping one row of marigold/25 rows of tomato.
3. Chlorothalonil spray against early blight and mancozeb against late blight.
4. 3 sprays of Ha-NPV at 28, 35 and 92 DAT.

This resulted in cost benefit ratio of 1:10 in favour of IPM against non-IPM and the pesticide reduction from 14 to 6 sprays.

IPM for Cucurbit Fruit Fly

1. Sowing resistant cultivars, lower cucurbitacin content offers resistance to pumpkin beetle. Thick and tough fruit rind offers low oviposition by fruit fly.
2. Bait prepared by macerated ripe banana fruit added with carbofuran and yeast to be placed at various spots in the field to attract and kill adult flies. Methyl euginol 2 per cent attracts males only.
3. Interculture to expose pupae by winter and summer digging.
4. Regular disposal of affected fruits to burry in soil.

5. Spray of NSKE 5 per cent to alleviate oviposition.
6. Use of small fruit size variety of bitter gourd as trap crop and insecticide application on trap crop to kill flies.
7. Clean cultivation to remove nearby bushy plants which provide shelter to male flies.

IPM in Mango (NCIPM, 2001)

Key Insect Pests

1. Hoppers *(Idioscopus* spp.*)*
2. Mealy bug *(Drosicha mangiferae)*
3. Inflorescence midge *(Erosomyia indica)*
4. Leaf webber *(Orthaga cuadrusalis)*
5. fruit fly *(Bactrocera dorsalis, B. zonatus)*

Key Diseases

1. Powdery mildew *(Oidium mangiferae)*
2. Die back *(Lasiodiplodia theobromae)*
3. Gummosis *(Lasiodiplodia theobromae)*
4. Anthroenose *(Colletotrichum gloeosporioides)*
5. Sooty mould *(Capnodium* sp.*)*

IPM Calendar

January: Cleaning of alkathene band, spraying fenitrothion (0.05%) for inflorescence midge and removal of weeds and infected young leaf to control powdery mildew.

February: First spraying of NSKE (5%) to control hoppers if noticed @ more than 5/panicle.

March: Second spraying of NSKE at pea size fruit stage and first spray of Sulphur (2 g/1) for powdery mildew.

April: Third spray of NSKE if required after a fortnight from second spray, second spray of Sulphur and removal of leaves infested with powdery mildew and malformed panicles.

May: Hanging methyl euginol traps (0.1%) + malathion (0.1%) to control fruit fly.

June: Methyl euginol traps to continue, early harvesting of mature fruits and collection and destruction of infested fruits.

July: Deep ploughing of orchard to expose eggs and pupae of mealy bug and inflorescencemidge and application of round up at the rate of 10 ml/1 water to control weeds.

August- Sept.: Removal and disposal of weeds formed by leaf webber, spraying with a safer insecticide if infestation noticed and pruning of over crowded and overlapping branches.

October: Flooding of orchard to control to mealy bug eggs, pupae of midge and fruit fly, pruning of infected and dried branches from 10 cm below and pasting of copper oxychloride to control die back, spraying 0.3 per cent copper oxychloride after pruning to control die back, phoma blight, anthracnose and red rust disease, removal of diseased foliage/twigs infected with anthracnose (twig blight phase) and removal of weeds.

November: Deep ploughing to expose eggs and pupae of insects and weeds, second spraying of copper oxychloride (3 g/l water) to control die back and foliar diseases and collection and burning of diseased and chopped leaves.

December: Fastening of alkathene sheets of 400 gauge thickness 25 cm wide around the base of trees to control mealy bug and raking soil around tree trunk and mixing with neem cake to control mealy bug nymphs. Applications of *Beauvaria bassiana* around tree trunk to manage nymphs of mealy bug.

IPM for Rhinocerus Beetle on Coconut (Pillai *et al.*, 1993)

1. Planting legumes as cover crop to conceal the potential breeding sites.
2. Treatment of breeding places with *Metarhizium anisopliac.*
3. Promote *Baculo virus* spread by leaving some dead standing palm.
4. Release of infected beetle to increase virus prevalence.

Insecticides Based IPM

Insecticides are in no way inferior to antibiotics if used selectively in proper dose, time and mode of application. Till date, these are the only component to combat certain specific problems. These can be used for seed treatment and seedling dips to provide safety during germination and initial stage infestations. Similarly, these may find place in IPM for selective root zone placement, whorl application, patch treatment, electrostatic spraying, chemigation, injection, infusion or localized systemic treatments.

Insecticides from sole resort to prepare poison baits particularly against fruit fly and grasshopper complex. The success of nipping the congregated population of migrating pests in the bud can be met only with insecticides. They form an integral spray device to control temperate fruits as dormant sprays. In situations, where a few pest number leads to devastating effects during early and non-consumption stage of crops, like whitefly transmitting viral diseases in fruits and vegetables may be controlled by quick knock down agents. This does not however emphasize to rely on these chemicals in principle. There is priority need to develop management strategies with ability to eliminate insecticides from agricultural system. The alternative residual sprays as ovicides, anti-internal plant tissue feeders and quick mortality causing substances have to be developed to replace insecticides in larger proportions.

Limitation of Existing IPM

Although, most of the IPM packages are eco-safe require low cost inputs, these are applied repeatedly in a crop season/year due to the lack of self-perpetuation ability and recycling. Further, these have to be given coverage on plot-to-plot scale involving initiation and resources to be met by farmers for a specific ecology at a time. These are not capable of maintaining steady state population dynamics for a longer duration and do not stand on the multiple criteria of sustainability as a part of integrated crop management system (Dhaliwal and Arora, 1994).

Essentials for Sustainability

In order to make IPM practices more sustainable, it is imperative to adopt management strategies extended to fauna and flora of cultivated as well as other linked ecologies in spatial scale of biogeographical regions in place of existing field-to-field considerations. The strategies should have regeneration potential on long term basis to maintain a steady state or oscillating dynamics of populations and thus culminating the need of their season to season application. The criterion of treatment applications should not be restricted to existing ETL approaches and should be enlarged to multiple criteria of sustainability. To achieve above mentioned objectives, it is indispensable to design an overall balanced ecosystem to minimize pest outbreaks rather than mere integration of compatible strategies being used presently.

Important Components of Sustainable IPM

Insect pest management strategies being used presently on vegetables, fruits, spices, ornamental and plantation crops do not have much scope and promise to improve sustainability actively. These may specifically be selected either to prevent losses of existing sustainability on one hand and act as supplementary measures in overall framework of integrated crop management schedules. These objectives may be achieved by using profusely the following strategies.

Conservation of Natural Enemies of Pests

The stable balance in faunal population in horticultura ecosystem may be promoted by prohibiting pesticide applications These toxic substances cause more damage to parasitoid pesticide and predators as compared to target pest species The naturally occurring beneficial fauna in this ecosystem ha been reported to be very high sufficient to check pest outbreaks. The important insect natural enemies of majo pests of important vegetable and fruit crops have been show in Table 2. Each pest species is subjected to the attacked b a number of natural enemies, if the latter are protected fro harmful substances and food scarcity. Poly crop and mix cro systems tend to provide stable shelter food and breeding medi on the diverse group of available host insects (Singh, 1999

One of the important dimensions of encouragement of useful fauna is to make use of favourable factors in interaction of these bioagents with crops *vis-a-vis*, the associated pests, *Trichogramma* sp. does not parasitise eggs of insects on chickpea, pigeonpea or tobacco but does it effectively on tomato, cowpea, marigold and many other crops and weed hosts.

Effective Bioagents and Biopesticides to Manage Pests of Horticultural Crops

Virus

Baculovirus infected coconut rhinoceros beetle *(Oryctes rhinoceros)* are released @ 10 beetles/ha giving areawise coverage at middle of the field at night to infect healthy beetles infesting coconut. Granulosis virus are sprayed to control codling moth, *Cydia pomonella* in apple orchards at the rate of 8 × 107 virus capsules/ml thrice at fortnightly interval in the afternoon. The first spray is done a week after first release of *T embryophagtim.* Nuclear polyhedrosis virus specific to *Helicoverpa armigera* (Ha-NPV) and *Spodoptera litura* (Sl-NPV) are very effective on tomato and a number of vegetable hosts by spraying homogenate of infected larvae @ 250 *l.e.* / ha during evening hours to protect from ultraviolet radiation.

Bacteria

Bacillus thuringiensis preparations under several forumulations including Bt oligos porogen mutants (300 g) are sprayed with knap sack sprayer at weekly interval as per need to control *Plutella xylostella* on cabbage and cauliflower. The bacterial preparation needs to be mixed with permitted spreaders and the spraying is organized afternoon. *B.t.* var. kurstaki is also sprayed at the rate of 0.5% a.i. to control *Papilio demoleus,* the lemon butterfly, in citrus nursery and young orchards. Normally, a single application is needed in each generation. This biopesticide is available in market under different trade names and forms an effective treatment to control *Helicoverpa armigera* larvae and a number of other lepidopteran pests on various crops. The gene producing the toxin in insects has given wider scope of its transfer in many vegetable and other crops to confer resistance in host crop to *H. armigera.*

TABLE 2
Natural Enemies of Pests in some Horticultural Crops

Crop	*Some important pest species*		*Approx. No. of natural enemies*	*Important enemies*	*Natural control efficacy*
	Common name	*Scientific name*			
Potato	Peach aphid	*Myzus persicae*	17	*Aphelinus* sp.	100% at Simla, 70% at Bangalore
	Epilachna beetle	*Epilachna vigintioctopunctata*			
		E. ocellata	8	*Pediobius foveolatus*	77%
	Cut worm	*Agrotis segetum A. ipsilon*	7	*Macrocentrus collaris*	Very high at Simla & Kodagu (Karnataka)
			5	*Cotesia reficrus*	High in plains of India.
	Tuber moth	*Phthorimaea operculella*	27	*Bracon gelechiae*	30-33%
	White grub	*Anomala* spp.	24	*Bacillus papillae*	Very high
Tomato	Fruit borer	*Helicoverpa armigera*	30	*Completis chloridae*	Upto 60%
				Trichogramma spp.	Upto 80%
Brinjal	Shoot & Fruit	*Leucinodes arbonalis*	8	*Eriborus* sp. *Phanerotoma* sp.	Very high
	Borer Brinjal	*Henosepilachma vigintiocto-punctata*	8	*Pediobius foveolatus*	30-60%
	Beetle Mealy bug	*Coccidohystrix insotata*	15	*Leptomastrix nigrocoxalis*	Very effective
	Red spider mite	*Tetranychus cinnabasinus*	9	*Cunaxa setinostrix*	
		Brevipalpus californis		*Amblysieus multidentatus*	Predatory effective mites

Chillies	Peach aphid	*B. phoenicis*	12	*Aphelinus* sp.	Upto 94%
		Myzus persicae		*Aphidius* sp.	Upto 97%
Crucifies	Diamond back moth	*Plutella xylostella*	29	*Cotesia plutellae* (Larvae)	Most effective (Karnataka & Solan)
				Diadegma fenestrale (do)	High altitudes
				Macromalon orientale (do)	Shilong
				Oomyzus sokolowskii (Pupae)	32-70% pupae at Banglore & 18-68% at Coimbatore
				Brachymeria excarinata	60% pupae
	Cabbage web worm	*Crocidolomia binotalis*	10	*Palexorista salennis*	Key parasitoid
	Cabbage borer	*Hellula undalis*	4	*Bracon* sp. (Larvae)	17% in spring
				Mermithid nematode	23% in August
	Cabbage butter fly	*Pieris brassicae*	10	*Cotesia glomerata* (Larvae)	65%
				Hyposoter ebeninus (Larvae)	16%
	Linseed semilocper	*Trysanoplusia orichalcia*	7	*Cotesia ruficrus* (Larvae)	Very high
	Aphids	*Lipaphis erysimi*	20	*Ehpeodes* (Syrphid)	Effective in March
		Brevicoryne brassicae		*Diaeretiella rapae* (Braconid	do
		Myzus persicae		*Various* spp. of coccinellids	
Okra	Fruit borer	*Earias* spp.	—	*Rogas aligarensis*	—
				Bracon hebetor, *Agathis* sp.	
				Trichogramma sp.	

(contd.)

Crop	*Some important pest species*		*Approx. No. of natural enemies*	*Important enemies*	*Natural control efficacy*
	Common name	*Scientific name*			
	Jassid	*Amrasca biguttula*	—	*Anagrus empoascae*	—
				Stethynium empoasca	
Mango	Leaf hoppers	*Idioscopus clypealis*	23	*Verticillium lecanii*	Key fungal agent
	Mealy bug	*Rastrococcus iceryoides*	23	*Anagyrus* nv. *Dactylopii*	Key fungal agent
		R. invadens		*A. mangicola*	do
	Leaf webber	*Orthaga euadrusalis*	8	*Goniozus* sp.	Main parasitoid in Kerala
	Diaspine scale	*Aulacaspis tubercularis*	8	*Pteroptrix koebelei*	Upto 97%
	Leaf gall midge	*Procontarinia matteiana*	5	*Chrysonotomyia* spp.	Key parasitoids
				Tetrastichus spp.	
	Eriophyid mite	*Cibaberoptus kenyae*	—	*Amblyseius* sp.	6.21% leaf in June
	Slug caterpillar	*Parasa lepida*	—	*Eocanthecona furcellata*	Key larval predator
Citrus	Citrus psylla	*Diaphorina citri*	15	*Tamarixia radiata*	Important nymphal parasitoid
				Diaphorencyrtus aligharensis	(upto 95%)
	Leaf miner	*Phyllocnistis citrella*	12	*Ageniapsis* sp.	80% in Punjab
	Lemon butterfly	*Papilio demoleus*	16	*Telenomous* sp. nr.	78% eggs
				Incommodus	
				Melalophacharops sp.	Key larval parasiroid
				Distatrix papipionis	do

	Blackfly	*Aleurocanthus woglumi*	16	*Encarsia divergens*	90% in Gauhati
				E. merceta	80-90% in Pune & Bombay
	Whitefly	*Dialeurodes citri*	11	*Encarsa lahorensis*	Key parasitoid
	Green scale	*Coccus viridis*	23	*Coccophagus ceroplastae*	43% in July
				Encyrtus lecaniorum	
	Yellow scale	*Aonidiella citrina*	4	*Pharoscymnus horni*	Promising predator
	Red scale	*Aonidiella aurantii*	5	*Comperiella bifasciata*	Main parasitoid
				Aphytis melinus	
	Fruit sucking moth	*Achoea janata*	24	*Trichogramma chelonis*	Egg parasitoid
	Leaf roller	*Psorosticha ziziphi*	6	*Goniozus* sp.	Key parasitoid
				Ophion triangularemaculatus	
	Mealy bug	*Planococcus citri*	10	*Coccidoxenoides peregrinus*	30% parasitism
		P. lilacinus	8	*Tetracnemoidea indica*	Key parasitoid
		Nipaecoccus viridis	6	*Anagyrus dactylopii*	Upto 90
Ber/	Aphid	*Aphis gossypii*	5	*Coccinellids and Syrphids*	Complete check
Guava	Mealy bug	*Ferresia virgata*	9	*Belepyrus insularis*	Key suppressing agents
				Spalgis epius	
	Green shield scale	*Chloropulvinaria psidii*	9	*Coccophagus ceroplastae*	Key suppressing agents
	Wax scale	Drepanococeis chiton	10	*Anicetus ceylonensis*	Key suppressing agents
				Chilocorus nigritus	

(contd.)

Crop	*Some important pest species*		*Approx. No. of natural enemies*	*Important enemies*	*Natural control efficacy*
	Common name	*Scientific name*			
Grape	Mealy bug	*Maconellicoccus*	14	*Anagyrus dactylopii*	62% parasitism
	Leaf roller	*Sylepta lucanis*	—	*Apanteles clitra*	Larval parasite
				Cardiochiles fulous	
Pomegranate	White ash white fly	*Siphoninus phillyreae*	—	*Encarsia azimi*	Suppresses pupae
	Aphid	*Aphis punicae*	—	*Scymnus* sp.	do
	Butterfly	*Deudorix epijarbas*	6	*Ooencyrtus papilionis*	70% parasitism at Rahuri
	Mealy bug	*Planococcus lilacinus*	5	*Spalgis epius*	Effective control
				Cryptolaemus montrouzieri	
Sapota	Bud & fruit borer	*Nephopteryx eugraphella*	5	*Xanthopimpla* sp.	Larval parasite
Custard apple	Mealy bug	*Maconellicoccus hirsutus*	—	*Spalgis epius*	Suppress population
		Ferresia virgata		*Cryptolaemus montrouzieri*	
Peach	Leaf curl aphid	*Brachycaudus helichrysi*	14	*Episyrphus boltaetus*	Suppress population
	Leaf gall midge	*Phyllodiplosis jujubae*	—	*Systasis* sp.	37-65% larval parasitism

Insect Parasites and Pr. dators

Trichogramma embryo at the rate of 2000 adults/tree is released at weekly interval to manage the menace of codling moth, *Cydia pomonella* on apple and other temperate fruits. First release is done once the first moth is caught phagum in pheromone trap per 10 tree or notice of egg laying. The application is continued till 5 months/trap/week are observed. *Trichogramma brasiliense* is also released to control: *H armigera* on tomato at the rate of 50,000/ha as adults or through stapling parasitised egg cards at 5 weekly interval 6 times from 25th day after transplanting or during oviposition duration. First release is recommended as soon as the first moth is caught in pheromone trap or egg noticed.

Cryptolaemus montrouzieri, this beetle has been extensively used to control a number of pests including citrus mealy bug, *Planococcus citri* by releasing 10 adult beetles/infested plant after initiation of blossom. Simultaneously suppression of ant is needed by ringing the tree base at one feet radius in soil by 5 per cent diazinon or 3 per cent carbofuran granule for better efficacy of the beetle. On guava, this beetle controls green shield scale, *Chloropulvinaria psidii* by releasing 10-20 adults/infested plant as soon as the infestation is noticed. Only one release causes suppression of scale due to its preference for guava ecosystem. For grape mealy bug, *Maconellicoccus hirsutus,* the beetles at the rate of 2500-3000/ha or 10 beetles/ infested vine after start of pests infestation, yields good results. This beetle has also been recommended for endemic areas against mealy bug, *Coccidohystrix insulata* on brinjal at the rate of 2 per plant. Protection from ant is necessary for better biological control efficacy.

Chilocorus spp. *Chilocorus bijugus* at the rate of 20 adults or 50 neonate grubs/infested tree is recommended for release in apple orchards to control sanjose scale, *Quadraspidiotus perniciosus* once in spring and to be repeated in endemic areas. C.nigrita controls Aonidiella aurantii by single application/ generation at the rate of 15 adults/tree. The ant suppression is necessary by basal ring application of a systemic granule for better efficacy.

Rodolia cardinalis, the *vedalia* beetle provides very good protection on citrus from *Icerya purchasi* at the rate of 10 adults/infested plant when released after notice of infestation. Simultaneous care to suppress ant population is necessary.

Aphelinus mali when released at the rate of 1000 adults or mummies placed soon after start of infestation of wooly aphid, *Eriosoma Janigerum* is reported to be very effective treatment. This parasitoid is effective against aerial population of pest and is more effective in valley than on slopes.

Encarsia perncıosi takes good toll of sanjose scale on apple through release of 2000 adults/infested tree once in spring and to be repeated in endemic areas.

Leptomastrix dactylopii controls citrus mealy bug, *Planococcus citri,* at the rate of 5000 adults on need base under expert supervision with simultaneous suppression of ant population.

Goniozus nephantidis is very promising against *Opisina arenosella* on coconut by need based release of 3000 adults/ pest generation. If the other bioagents like *Elasmus nephantidis* and *Brachymeria nosatioi* are also present along with G. *nephantidis,* the parasitoid host ratio is 2:5 in coconut ecosystem, no action is required. The exclusive required ratio of G. nephantidis and host should be 1:5.

Mite

Spider mite, *Tetranychus spp.* on beans are effectively controlled by *Phytosciulus persimilis* by releasing 10 adults/ plant 30 days after germination or when infestation is noticed. This predatory mite is also effective on Tetrancychid mites on straw berries brinjal and some other vegetable crops.

Fungus

Verticillium lecanii has been found to be very effective to suppress mealy bugs and scale insects. Spraying spores at the rate of 106 spores/ml along with 0.005% quinalphos by knapsack sprayers ensuring proper coverage in afternoon is very effective even by single application just before the onset of rains checks the infestation of green scale, coccus viridis on citrus. Addition of permitted spreader in spray is necessary. *Aphis gossypii* on guava is effectively controlled by application of 109 spores of

V.lecanii along with 0.1 per cent teepol as surfactants. The first spray is alone before the onset of rainy season followed by need base applications.

Clean cultivation and sanitation

Since many species breed or shelter during off season or undergo stage transparent on various plant debris, removal of dead leaves, stems, roots, twigs stubble and fruits from their habitat reduces chances of their survival, continuity and further dispersal. Regular collection and destruction of fallen leaves and fallen fruits of apple is the essential measure to reduce the population of codling moth. Egg parasitoids released in clean orchards are more effective than uneared ones. For borers complex on temperate fruits viz. apple root borer, flat headed stem borer, shoot and trunk borer, sanitation, pruning and disposal of dead heart, intercultural etc. from the essential steps for pest management besides nutrients management. This is true for sanjose scale also (Tandon, 1994).

Regular removal and destruction of fallen fruits of mango, citrus and guava reduces the fruit fly population in the orchard and stone weevil incidence in case of mango. To reduce the menace of these pests, it is desirable to prune and remove shoot borer infested twigs in March-April and stem borer infested twigs in August on a community basis. The population of leaf miner, scale insects, mealy bugs, aphids and psylla can be reduced by removal of water shoots in citrus. Conservation of natural enemies of fruit fly and other parasitoids in soil can be encouraged by chisel ploughing.

A number of pests prefer and seek shelters away from the main plants to nearby habitats. Cucurbit fruit fly uses nearby habitats. Cucurbit fruit fly uses nearby bushes to hide in large numbers. Citrus mites are more abundant on plants situated on the road sides under dust cover. Removal of shelting bushes and sprinkler washing of dusted leaves prevent free breeding of these pests in the surroundings.

Neem, an Ideal Product in IPM

Neem offers a miraculous solution to control viral problems in horticultural crops (Singh and Singh, 2002). Its aqueous leaf extracts caused up to 87 per cent inhibition against PVX on tomato plants and seed kernel extract against apple mosaic ilavirus on cucumber. Neem oil (1%) spray causes significant reduction of ring mosaic desease of peanut caused by SWV. A mixture of neem, karanj, castor, mahua and sesame is used by farmers to control or repel insect vectors of viruses. Virus vectors like *Aphis gossypii* on okra, *Acyrthosiphon pisum, Aphis fabae* and *Myzus persicae* are effectively controlled by NSKE or neem cake. Some commercial products like Margocide-OK cause inhibition of viral transmission by *Myzus persicae* and whitefly, *Bemisia tabaci.*

Aqueous leaf extract is reported to control bacterial leaf spot of chilli caused by *Xanthomonas compestris* pv. Vesicatoria. The cucurbit wilt transmitted by bacteriae, Acalymoma and Diabrotica duodecimpunctata can be effectively controlled by neem causes damping off, root rot and crown rot in many crops. Neem planted along with tomato, brinjal, cabbage and cauliflower shows reduction in stylet bearing nematodes population.

As insecticide, commercial products of neem like Margosan-O and Neemazal-S reduce setting activity in apterous adults of *Myzus persicae* and bean aphid, *Aphis fabae.* Egg laying has been reported to decrease in *Bemisia tabaci* on tomato and *Dysdercus fasciatus* by NSKE on okra. Deterrence to egg laying by neem oil treatment has been recorded in cabbage webworm, Crocidolomia avonana, bollworm, Helicoverpa armigera and a number of dipterous insects. It shows antifeedant effect against many lepidopterous larvae. The antihormonal effect of neem causes reduction in fecundity of pea aphid, *Spodoptera littoralis* and *Plutella xylostella.*

Host Crops-pest Resistance and Interaction Mechanism

More than 200 cultivars of fruits and vegetables with resistance to 40-50 insect spp. have been released and are used in commercial production in USA (Wiseman, 1990). In tomato a number of open pollinated varieties are grown under rainfed conditions in large scale. Among bybrid varieties Arka Vardhan is bacterial wilt and nematode resistant, Avinash-2 is popular hybrid tolerant to leaf curl. Region specific disease resistant varieties are available for cabbage, okra and also all vegetable and fruit crops. Good success has been achieved in developing resistance to diseases. However, no tenable genetic resistance to pests in horticultural crops is available. Some resistant varieties show reduced damage and lesser the number of insecticide sprays. The existing variation in cultivars in respect to maturity groups, harvesting time, morphological differences have been used successfully for manipulation of pesticide resistance. Screening of germplasms is useful in resistance breeding programmes with biotechnological tools available now. However, no crop has varieties with gene conferring pest resistance. In solanaceous vegetables, growing of tomato, potato and brinjal regularly in the same field makes the soil sick with bacterial deseases and knot nematodes.

Studies on insect host interaction mechanism have given a new dimension in designing pest resistant crops. Metcalf *et al.* (1982) identified atleast 20 different chemicals from different parts of cucurbitacae. Oxygenated tetracyclic triterpenes commonly called cucurbitacins A to S function as defence against a number of herbivores as feeding similarly 100 chemicals from glucosinolates like isothiocyanate, allylisothiocyanates and sinigrin in cruciferous and other plants affecting host preference. The terpenoids like drimane inhibits feeding in spodoptera spp. and *Myzus persicae* by inhibition of gustatory stimulus (Asakawa *et al.,* 1988). Limonoids like azadirachtin from neem is the most promising pesticide as feeding deterrent to more than 100 insect spp. (Saxena, 1989). It is also toxic and shows antibiosis with pronounced morphological deformities. High concentration of rutin, a flavonoid, and chlorogenic acid is present in leaf trichomes of tomato which are toxic to *Heliothis zea* (Isman and Duffey, 1982). Antiherbivory functions of coumarins due to their toxicity against virus to vegetables are ovicidal to potato beetle. Isoflavones coumestrol is antifeedent and antibiotic to cabbage looper. Biotechnological approaches have made it possible to transfer Bt genes in tomato, potato and a number of crops imparting high level of resistance to insects under field conditions (Hilder *et al.,* 1987). Molecular genetic maps are now available in these crops. Rotenone in this group is a well known insecticide. The pyrrolifine alkaloid DMDP is very potent digestion inhibitor to *Callosobruchus maculates* (Evans *et al.,* 1985). Many such combined factors of resistance in solanaceous and leguminosious vegetables have been recorded (Panda and Khus, 1995).

Biorationals

Utilization of chemicals affecting insect behaviour, growth or reproduction and population suppression viz. chitin synthesis inhibitors, anti-hormone, pheromones and allelochemicals may be done for many pests of horticultural crops (Dhaliwal and Arora, 1998). There is need to test their efficacy against problem pests with standardized methodologies and economic considerations.

Future Research Priorities for Sustainable Pest Management

Agro Ecosystem Analysis

A typical agro-ecosystem, largely created and maintained to satisfy human needs, is characterized by interactions among almost uniform plant population, weed, animal and micro biotic communities and the physical environment. Ecological succession in this system is never allowed to reach climax and terminates with crop harvesting particularly in annual crops. The energy flow across the food web in such case is hardly cyclic and is prone to wider fluctuations. Monoculture, islands surrounded by other crops, grasslands or fallow fields undergo unstable colonization accommodates only a limited number of species exhibiting large number of progenies and high dispersability thus creating vacant sites to be occupied by others. This results in imbalance in ratios among different food chain levels and leads to outbreak of specific population. Since the ecological niche. i.e. functional position of an organism in the community, cannot be identical at a specific place and time, the uniform and simple ecosystem witness interspecific competition. Thus, from the basic ecological considerations of ecological succession, energy flow, island colonization and existing niche, the horticultural ecosystem is not sustainable primarily for want of biodiversity and a larger biogeographical dimension.

Studies on the basic ecological interplay on balanced system is necessary. Even in huge and megadiverse and forest ecosystem, intra-and-specific competitive forces lead to evolutionary adaptations both in plants and animals. Insects have excelled in this long physical, chemical and behavioral warfare by vivid acquisitions and adaptations for host selection on one hand and protection from changing environment on the other. The bionomics of pest through seasons have indicated weak links to manage their population at most easily approachable stage. Recent works on ecophysiology of Indian gypsy moth has given very vital information for its management. Evidences of migration in many insect spp. Including *Helicoverpa armigera* may help solving this great problem in line with locust control. Indepth studies on interplay of pests with environment will afford precise forecasting models.

Biodiversity Management

Agro ecosystem tends to be more susceptible to host damage and prone to outbreaks in lack of diversity in plant and insect species and its sudden alternation endowed by weather, and man. Agricultural system with lower biodiversity as compared to forests and natural ecosystems, harbours only a few major species and numerous minor species. Thus, the pest problems are directly related to agronomic practices. Monoculture and overlapping cropping are more vulnerable to pest outbreaks.

Poly crop system favours rich culture of parasites and predators to attack various stages of pests and prevent severe pest outbreak. Crop mixes and variety mixes make the agro ecosystem more sustainable.

The role of entomophagus insects and pathogens in regulation of population of forest defoliator, gypsy moth, *Porthetria dispar* and brown tail moth, Euproctis chrysorrhoea was emphasized by Clark *et al.,* 1967. High yielding varieties, crop monoculture, extensive use of fertilizers and pesticides, irrigation and change in crop spacing and rotation have largely contributed in dominance of only a few species in faunal population.

Thus, one of the means to control insects is to create deterrence of their colonization through intra-field diversity by intercropping, trap cropping, weed culture, adjacent cropping. Tomato intercropped with cabbage (1:1), 30 days later to tomato reduces the diamond back moth on cabbage. Allelopathic effect of taramira in raya is reported to decrease population of mustard aphid. Rows of marigold or tobacco in tomato crop divert Helicoverpa and also helps colonization of some beneficial fauna. Okra or castor around cotton field as trap crop diverts the jassid, bollworm and Spodoptera population on border crops with a view to balance the stability and productivity, combined agro-forestry and agro-horticulture system may be adopted to impart biodiversity in overall system.

Pest Resistance Management

Development of adaptation in insects to any human health or plant protection mechanism including chemicals, cultural, biological or biotechnological control are common in situations of repeated and long term continuation of any control device. This occurs through genetic selection as a result of increased selection pressure. More than 600 insect spp. are reported globally to develop resistance against pesticides making them for non *effeicacious*? in insect control. In India alone, 16 insects belonging to human health 13 of stored products and 8 from agricultural crops are now resistant to various group of insecticides. Four of 8 agricultural pests belong to horticultural crops namely DBM, tomato fruit borer, while fly and brinjal fruit and shoot borer and tabacoo caterpillar in addition to red spider mite (Mebrotra, 1991; Singh *et al.*, 1994; Mukherjee and Singh, 2003).

This resistance phenomena is not restricted to insecticides. There are numerous reports showing its development to cultural, biological, biopesticidal, transgenic measures and towards a number of pheromones, crop rotations, sowing dates, attractants and repellents, which form the main IPM components in present form. Such resistance has also been documented toxicants in bacteria, weeds, human pathogens, insects, mites and rodents. Invariably these situations lead to reduced efficacy of strategies due to continuous and repeated use of single control measure. This envisages attention to monitor the efficacy of various employed methods used on long term basis and necessary steps to evolve proper pest resistance management strategies as component of IPM.

The five principles emphasized and used effectively to contain or reduce resistance are (1) application of diverse mortality mechanisms oriented strategies, (2) reduce selection pressure of individual measure, (3) management of susceptibility in pest population including refusia management, (4) regular monitoring of resistance under field conditions and (5) establishment of policy and communication. Good progress in India has been made to reduce insecticidal resistance in fruit borer, *Helicoverpa armigera* on cotton under multilocational ICAR/NRI/ICRISAT supported projects.

Global Opportunities

The World Trade Agreement (WTA) provides excellent opportunities for exports of more farm produce. This would need attention to both sanitary and phytosanitary measures. Tropical fruits and flowers, processed products and selected foodgrains have high export potential. We have to try to capitalize on existing strength of institutional support, technologies available, varying agroclimatic conditions, good land, water and relatively cheaper labour. Pesticide residues is likely to become a serious non tariff barrier. More thrust has to given to develop no insecticide based IPM for 411 export oriented crops. IPM technologies for green house and poly house culture will also be needed.

Wider Area Adoption of IPM

Considering 143 million hectares cultivable area with 110 million holding, the wide area adoption of IPM technology seems to be a Herculean task. For some crops like rice and cotton, IPM technologies have been adopted at village scale (Puri, 1996). More wider farm areas are to be clustered for sustainability at regional level. Supply of different quality inputs will be a key factor in this endeavour. Some organized sectors like pesticide industry may view IPM promotion as a threat to their industry. The chemicals by virtue of commercial availability become the. choicest and assured means for the farmers.

REFERENCES

Asakawa, Y., Dawson, G.W., Griffiths, D.C., Lallemand:, J.Y., Ley, S.V., Mon, K, Mudd, A., Pezechik-Leclaire, M., Pickett, J.A., Watanabe, H., Woodcock, C.M., Zhang, Z.N. 1988. Activity of drimane antifeedants and related compounds against aphids and comparative biological effects and chemical activity of (–) and (+) polygodial. *J. Chem. Ecol.,* 14: 1845-1855.

Clark, LR., Geier, P.W., Hughs, R.D. and Morris, R.F. 1967. *The Ecology of Insect population in Theory and Practice*. The English Language Book Society and Methuen & Co. Ltd., London, U.K.

Dhaliwal, G.S. and Arora, R. 1994. "Components of insect pest management : A critique". In: G.S. Dhaliwal and Ramesh Arora (eds). *Trends in Agricultural Insect Pest Management*, Commonwealth Publications, New Delhi, India, pp. 1-55.

Dhaliwal, G.S. and Arora, R. 1998. *Biorationals and other innovative approaches. Principles of Insect Pest Management*, Kalyani Publishers, Ludhiana, pp. 221-241.

Evans, S.V., Fellows, L.E., Shing, T.K.M. Fleet, G.W.J. 1985. Gycosidase inhibition by plant alkaloids which are structural analogues of monosaccharides. *Phytochem*., 24: 1953-1955.

Hilder. V.A.. Gatehouse, A.M.R., Shreeman, S.E., Baker, R.F., Boulter, D. 1987. A novel mechanism of insect resistance engineered into tobacco. *Nature*, 330: 160-163.

Isman, M.B. and Duffey, S.S. 1982. Toxicity of tomato phenolic compounds to the fruit worm, *Heliothis zea*. *Entomol. Exp. Appl*., 31: 370-376.

Krishna Moorthy, P.N. and Sardana, H.R. 2001. "IPM in cabbage and tomato". In: America Singh, T.P. Trivedi, H.R. Saradana, A. Dhandapani and Naved Sabir (eds), *Resource Manual on Validation and Promotion of IPM*. NCIPM, IARI, New Delhi, pp.90-100.

Mehrotra, K.N. 1991 . Current status of pesticide resistance in insect pests in India. *J. Insect Sci*., 4: 1-14.

Metcalf, R.L., Rhodes, A.M., Metacalf, RA., Ferguson, J., Metacalf, E.R. and Lu, P. 1982. Cucurbitacin contents and diabroticite (Coleoptera : Chrysomelidae) feeding upon *Cucurbita* spp. *Environ. Ent*., 11: 931-937.

Mukherjee, U. and Singh, H.N. 2003. Monotoring for Resistance to Fenvalerate and Monocrotophos in Diamondback moth, *Plutella xylostella* L. in Varanasi, Uttarpradesh, India. *Pesticide res. J*., 15(2): 200-204.

Nayar, K.K., Ananthakrishnan, T.N. and David, B.V. 1976. *General and Applied Entomology*. Tata McGraw-Hill Publishing Co. Ltd. New Delhi, pp. 469-557.

NCIPM, 2001. *Resource Manual on Validation and Promotion of IPM*. L.B.S. Centre for Agriculture and Biotechnology, IARI, New Delhi.

Panda, N. and Khus, G.S. 1995. *Host Plant Resistance to Insects*. CAB International and Wallingford, U.K., pp.151-206.

Pillai, G.B., Sathiamma, B. and Dangar, T.K. 1993. Integrated control of rhinoceros beetle. In M.K. Nair. H.H. Khan, P. Gopalsundaram and E.V.V. Bhaskara Rao (eds). *Advances in Coconut Research and Development*. Oxford and IBH Publishing Co. Pvt. Ltd., New Delhi, India, pp. 455-463.

Puri, S.N. 1996. Integrated Pest Management: An Entomological Approach to Sustainable Agriculture. *IPM and Sustain. Agni.*, 6: 1-6.

Saxena, R.C. 1989. "Insecticides from neem". In: *Insecticides of Plant Origin*, J.T. Arnason, B.J.R. Philogene and P.Morand (eds). *ACS Sysp. Series 387*. American Chemical Society. Washingto DC,. pp. 110-135.

Shukla, R.P., Mishra, A.K. and Sabir, N. 2001. Integrated Pest Management in Mango. In: America Singh, T.P. Trivedi, HR. Sardana, A. Dhandapani and Naved sabir (eds). *Resource Manual on Validation and Promotion of IPM*. NCIPM, L.B.S. Centre for Agriculture and Biotechnology, IARI, New Delhi, pp.101-105.

Singh, America, T.P. Trivedi, H.R. Sardana, A.Dhandapani and Naved Sabir (ebs). NCIPM, 2001. *Resource Manual on Validation and Promotion of IPM*. NCIPM, L.B.S. Centre for Agriculture and Biotechnology, IARI, New Delhi, pp. 20-38.

Singh, H.N., Rose, J. and Singh, H.K. 1994. Helicoverpa armigera insecticide resistance management at Varanasi. *Pod Borer Management News Letter No. 4*, p. 3.

CHAPTER 9

Sustainable Plasticulture in Agriculture

Bishnu Narayana Sethi*

Sustainable: Meaning and Introduction

Producing plenty of quality food, protecting and enhancing the soil, water and other natural resources building a thriving rural economy , giving farm families and communities a good life on the land - sustainable agriculture will do all these things for decades and centuries to come but sustainable agriculture does not yet exist - not in the sense of a model farm or alternative farm economy that can be pointed to and emulated.

Sustainability is a long-term goal for research and development for farm planning and the farmers daily work and for society in general. More and more people are making sustainability a guiding principle, they persistently put the question to system and tools both old and new can we use them without westing resources and stealing from future generations?

Modern high input agriculture has produced great increase in crop yield government policies have kept this abundance cheap for the U.S. consumer. But the social and environmental costs have been high. Loss of topsoil, water pollution loss of biodiversity, dependence on non renewable resources, rising production costs and falling prices for crops the decline of rural communities and fewer and fewer farmers are some of the unsustainable result of industrialized farming.

*Lecturer, Laxminarayana Degree College, Kodala-761032, Orissa.

Over the past several decades, farmers and researchers around the world have developed ecology-based production systems called by an array of names. Organic, natural, low input, integrated, alternative, regenerative holistic, biodynamic, biointensive and biological farming systems all seek sustainability. While none has arrived the answer, all have made important contributions and achieved impressive results on thousands of farms Ecology-based farming mimics nature to create an agro-ecology where biodiversity is high, plant nutrients are recycled soil is protected from erosion, water is conserved and not polluted, tillage is minimized and live stock production is integrated with perennial and annual crops. Natural resources and managed biological system from the foundation of farm production. But production has a social and economic context. For farmers threatened with extinction and desperate to survive, sustainability foremost meaning is economic solvency, a decent family income and good quality of life. Farming will not be sustainable until farmers reliable markets in which skill and hard work are fairly rewarded.

To Achieve Sustainability

Farmers and other agricultural thinkers have established a strong set of guiding principles based on stewardship and economic justice. Farmers and researchers are developing new practices and improving old ones. Sustainability is being manifested one crop, one field, one farm at a time conventional age and the land grant university system have also contributed. Alternative ideas have entered the mainstream and sustainability is a research and education priority in more and more institutions. Mean while consumers and grassroots activists are pushing for alternatives in the market place and changes in national farm policy. They are working to create direct connection between farmers and consumers to promot regional food self sufficiency, to reduce economic concentration in production, processing and marketing and to encourage resources conservation. Unlike those who dream of completely mechanized mega farms feeding the world. These futurist envision small to mid size diversified farms growing food primarily for local and regional consumption.

Plasticulture: Productivity and Sustainability

The use of plastics in agriculture (Plasticulture) is one of the recent and major developments for increasing the agriculture production. This technology has widely been accepted and implemented in many western countries and middle-east for higher and quality production. In India plasticulture is in very infant stage and has vast scope for improvement. The percapita consumption of plastics in India is about 2.5 kg which is very low compared to world average which is 15.4 kg and in some developed countries it is more then 50 kg. Therefore, it is no wonder that platiculture is being considered as one of the most important vehicles which could be instrumental in leading agriculture to evergreen revolution.

Looking into the international experience of increase in agriculture productivity through plasticulture techniques the Government of India has constituted a National Committee on Use of Plastic in Agriculture (NCPA) in l98l with a mandate to usher into technological break through in Indian agriculture by the use of plasticulture techniques since than more than 20 Plasticulture Development Centers (PDCs) are being established in different state agriculture universities and other researches institutes for effective research and development of technologies for maximizing the benefits of plastic at the end user. NCPAH was reconstituted in 1996 as the National Committee on Plasticulture Application in Horticulture (NCPAH). The NCPAH is headed by the union agriculture minister and its co-ordination cell is located at Delhi providing secretarial support for various activities related to plasticulture application. Research of plasticulture application in agriculture is being undertaken through ICAR Scheme. All India co-ordination research project on application of plastic in agriculture (APA) with its headquarter located at CIPHET, Ludhian. Besides this applied research work is being carried out through its 16 PDCs for standardization of different technologies. To be implemented at farmers field. In general, plastic has role in every field of agriculture but most important technologies are being described briefly in the article.

Canal Lining

Canal, like arteries of the human physical system, works as carries of water to the farmer's field for crop production.

A large quantity of water (35-45%) is lost through seepage in canal system and also causes water logging and salinity in the vicinity. The conversion of large fertile land into barren land due to water logging and salinity in north-western plains of India is the practical example of this fact. The experience indicates that lining of canal with plastic films is a convenient and economical proposition. The use of plastic films in canal lining was introduced in India as for back as 1959. More than 10000 km length of canal in different irrigation projects has been lined with LDPE films. The performance of these plastic film has been found very satisfactory. Similarly lining of ponds and reservoirs has also been tried out successfully. This prevents water losses due to percolation and there is tremendous scope in the years to come.

Plastic Pipes

Plastic pipes for conveying water, suction and delivery of agriculture pump sets have the maximum potential as it helps in reducing the evaporation and seepage losses, compared to open canals plastic pipes save at least 30-50 per cent of water loss, the use of plastic pipes is maximum Uttar Pradesh and few other states have also started implementing the same. Similar model is now being adopted by several other developing countries. Looking into the pros and cost of plastic pipes, presently scientific opinion favours increased use of plastic thus inviting industry to look for alternative to reduce cost of plastic pipes and durable man made materials.

Micro Irrigation Systems

Micro Irrigation System (MIS) is an irrigation which provides high frequency application of water in and around the root zone of plant system. The MIS consists of a net work of pipes along with a suitable emitting device. A typical MIS has drippers micro sprinklers, distribution line, control head system, fertilizer tank and fittings. The system has emerged as an appropriate water saving technique for widely spaced high value crops in water scarcity, undoubted sandy and hilly areas of the country.

Symca Blass an Israeli engineer has been considered as the father of the concept of drip/trickle irrigation. From a commercial beginning outside Israel in 1969 micro-irrigation in its current diverse form had been installed on some 11 million hectares throughout the world in 1997. In almost past two decades the growth of MIS in India has picked up momentum from 1500 hectares in 1983 to 0.3 million ha. The major advantages of MIS are saving in water improve productivity and quality of produce, reduce weed growth and disease incidence. The system also facilitate the use of fertilizers and pesticides along with irrigation water more economically. However, high investment at the initial stage and frequent clogging of emitters are some of its disadvantage. India is emerging as a leading country in the use of micro irrigation system. Research work being carried out at different plasticulture development centres located at different regions of the country has indicated that drip/trickle irrigation system can save water up to 84% with increased yield up to 60% besides improving the quantity of the produce.

Due to liberal support of the government, it has been possible to cover 0.3 million ha under micro irrigation system in the country on an average about 30,000 hectare area are being brought under MIS annually. However, against the potential of about 27 million ha the coverage so far is only 0.3 million ha. Therefore there is tremendous scope for rapid expansion of area under micro irrigation system.

Plastic Mulching

Plastic mulching is a practice of covering the soil surface around the plant to make conditions more conducive for plant growth through *in situ* moisture conservation, weed control better co exchange for root system and soil structure maintaince use of dry leaves, straw, hay etc. as a mulching material has been prevalent for ages. However, the introduction of plastic film as mulch increases the efficiency and hence increase in yield. LDPF and LLDPE plastic films are commonly used for mulching. LLDPE black plastic mulch films are more popular due to opacity which check the weed growth under the film. However, different colour polythene mulch have different effects on crop growth and pest management. In China, silver coloured film has been successfully used in cotten for better retention of soil moisture and ability of repelling aphid. In Israel, yellow colour mulch sheet is used for management of white-fly in tomato. In India extensive experiments have propped the enormous advantages of the use of plastic mulch.

The major limitation of use of plastic films is non-biodegradable nature and tends to pollute the environment. Hence it will be necessary to educate, the farmers for its safe disposal after use. Similarly frequent replacement is another problem as the life of plastic films are usually 12-20 months hence availability of plastic film at reasonable cost for replacement is also a limited factor.

Soil Solarization

Soil solarisation is a novel technique of controlling soil borne pest including weeds. It involves covering the wet soil with thin transparent ploythene films during the summer months. The process would raise the surface soil temperature by 8-12°C as compared to non-solarised soils. A duration of 4-6 weeks is sufficient to give satisfactory control of weeds and other soil borne pasts. The solarisation technique is simple and easy to use by farmers. However its immediate application appears to be more promising in nursery areas and in high value crops.

Green House

Green house technology has been found to be an appropriate intervention in the development of horticulture in India. Although the present area under green house in the country may be less than thousand hectares. There is a large untapped potential which would bring nutritional self-sufficiency besides enhancing the agri-exports and no farm employment.

Green houses are framed or inflated structures covered with transparent or translucent materials in which crops could be grown under partially controlled environment which is large enough to permit normal culture operation operations. Glass house, poly houses, walk in tunnels are all different names given to green houses depending upon the type of material used and the utility of the facility. The size of green house could vary from few meters to few hectares. Green house of longer size are usually constructed for export oriented project for vegetables and flowers. The growth of green house area in India in the recent years has been about 30 per cent per annum. The total green house area at present is estimated to be close to 1000 ha which is insignificant as compared to China, Japan and South Korea. Other countries where green house technology is widely used are USA, Netherlands, Israel, Canada, Spain beside some Arab countries.

Green houses are suitable for growing a variety of vegetables, fruits and flowers. The green houses help to do year round cultivation even under extreme climate conditions. In addition to temperature control, other benefits of green house cultivation include protection from wind soil warming, control of pests etc. Thus, green house cultivation could be considered as protected cultivation that enhances the maturity of crops, increases yield improve the quality of produces etc. The total time for preparation of seeding and cutting also get reduced significantly by the use of this technology.

High capital investment is one of the serious constraints is its wide adaptability. The package of practice for green house cultivation is yet to be standardized. There is urgent need for perfecting the agro-technique for cultivating inside green house. Besides green house other structure like low plastic tunnels shading, nets, antihail nets bird protection nets etc. are also miniature form of green houses to protect the plants from rains, winds, low temperature, frost and other vagaries of weather.

Plastic in Post-Harvest Management

Plastic have a major role to play in several post harvest operation like sun drying on black play there film protecting the produce on thrashing floors, drying yards and market places through temporary cover of plastics C A P storage of food grains where therc is shortage of godowns and plastic sandwich bins for an farm grain storage. These plastic storage bins arc very convenient and hygienic over the conventional muddy structures.

Similarly the subject of use of plastic in packaging of agriculture produces is very vast and diverse. It would cover a range of plastic products in one form on the other which could be useo for the packaging of different agriculture produce like food grain, jute cotton, oil seeds plantation crops like tea coffee, rubber, horticultural produce like fruits and vegetables, milk and milk products.

Sustainable Resource Management

Land, water, climate, flora and fauna are the basic natural resources for agricultures development which are subject to various kinds of deteriorating influences. Since agricultural development cannot subsist on a deteriorating natural base. It is imparative to develop strategies for conservation and improvement of resources. The concept of sustainable resources management implies that the need of the present can be met without compromising the ability of the resources to meet the need of the future. We have to emphasis on long term truly productive agricultural sustainability by way of sustenances of natural resources, economic viability and social acceptability of production system and protection of environment. The aim of eco-farming must be to match the crop soil and climate of a region maintaining the ecology infarming and gaining from the economy and efficiency of inputs. We have to reduce the burden of green house, gases like CO_2 and methane, globe warming and ozone depletion we need to follow the technology that reduce the demand on land, water and bio-diversity without adverse effects on agricultural production and nutritive value of food Nurse the soil back to health, change cropping patterns to maximize ecological productive, efficiency, improve water use efficiency through conjuetive use of rains, tanks, underground well and river waters.

Effective means to harvest the solar incidence both through photosynthetic pathway and other passive and active solar devices are to be included to meet the energy demands and increase the biomass production of the farms. Multi-tier cropping systems including legumes horticultural and silvi-pastoral species contribute in enhancing the overall productivity per unit of soil air and water.

To safeguard the environment from degradation and to maintain the purity of air, water and food. We have to opt for less and less use of chemicals and sift from chemical to ecological agriculture or nature farming, build up strong organic base to fertilize our fields. The recent energy crisis and the consequent price hike of fertilizer due to withdrawal of subsidy unfertilizer coupled with low purchasing power of the farming community have again reviewed interest in organic recycling throughout the world. Proper agricultural policies supporting the disserunation of the environment friendly and cost effective extension machinery encouraging progressive farmers to go a long way in ensuring eco-friendly sustainable agriculture in India.

Conclusion

The Plasticulture techniques have tremendous scope in Indian agriculture. The results achieved at scientific farms and laboratories have been found encouraging to higher crop yields, better quality and better post harvest management of produce. Compared to many developed countries plasticulture is in its primary stage in India. There is great need to create a awareness and disseminate the knowledge and development of plasticulture application in the country. Hence, to exploit the full potential of plasticuiture techniques. It calls for an integrated approach on the part of different agencies to bring about a plasticulture revolution with the strong backing of the government of India and private sector undertaking.

REFERENCES

1. APPU, P.S.C. (1996) *Land Reforms in India*, Vikas Publishing House, New Delhi.
2. Arneja, C.S. and Sandeepika Khara, *Kurukshetra,* July 2001.
3. Gupta, A.K. (1995) Rethinking policy options for watershed management, *Indian Farming.*
4. Singh, H. (1997) Sustainable development Under IWDP. Punjab - direct and indirect employment generated benefits.

CHAPTER 10

Ericulture: A Potential Income Generating Sector (Based on the Study Coducted in Orissa)

Jagadish Kumar Kar*

INTRODUCTION

Ericulture is a cottage industry of the poor. It provides gainful employment for the tribals, economically weaker sections and unemployed persons ensuring a sure sort return with less capital investments. It also provides opportunities of earning with less effort in a leisurely way staying at home for the idle manpower like house wives, old aged people and unemployed youths. Hence, ericulture can be designated as an important rural-based cottage industry so far its potentiality for income generation and employment generation is taken together.

Ericulture (rearing of eri silkmoth for erisilk) is a part of sericulture (rearing of silkmoths for silk). The productivity of ericulture largely depends on availability of host plant (food plant) species, type of soil on which the host plants are grown and the prevailing climatic and environmental conditions of the geographical area. The quantitative and qualitative production of eri silk depends mostly on types of host plant or food plant selected and the nutritional contents of the host plant leaves as well as the seasonal environmental parameters of the rearing environment. The larval stage is only the feeding stage in the life-cycle of eri silkmoth. Generally, eri larvae are polyphagus (feed on a wide variety of food plants). Castor

*Lecturer, Department of Zoology, Kshetrabasi Dav College, Nirakarpur, Dist. Khurda, Orissa-752019.

is the primary or principal host plant and most of the ericulturists and eri silkmoth rearers seem to rearer eri silkmoths on castor leaves. It is also accepted that eri silkmoth larvae reared on castor leaves produce comparatively better silks in quality. Generally, castor is cultivated as an agriculture crop or a mixed crop or a garden crop in India. However, large populations of wild varieties are found grown on the hills and plains through out the country. Castor bean (seed) is the product harvested from castor crop and it has a high commercial value for its oil content. On the other hand, the leaves are used to feed the eri silkmoth larvae, where eri silk is the final product.

Nowadays, eri silk has high demand in the local, national and international market owing to its multifarious utility for human happiness and comfort. Thus, ericulture in its productivity is a source for dual income provider, one source being from the castor bean and other from the eri cocoon or eri silk.

Art of Ericulture and Eri Silk Production

The art of eniculture and erisilk production involves all aspects of sericulture that finally leads to the production of fabrics. Primarily, it involves -

- Rising of host plants
- Production of eri silkmoth eggs
- Rearing of larvae till formation of cocoons
- Processing of cocoon to obtain silk yarn
- Utilization of silk yarn in weaving of silk fabrics.

The life cycle of an eri silkmoth *(Samia ricini* Donavan) passes through four distinct stages viz., egg, larva (caterpillar), pupa and moth (adult). At the beginning of the life cycle moths emerging from the diapausing (period of suspended animation and growth) cocoons mate to lay eggs. After about 7 to 15 days (depending upon the seasonal climatic parameters) of incubation, tiny larva comes out from the egg. The larval stage is the only feeding stage and comprises of five instars. The larva feeds on the leaves of a wide variety of host plants, namely castor *(Ricinus communis),* Kesseru *(Heteropanx fragrans),* Payam *(Evodia flaxinifolia),* Tapioca *(Monihot utilissima),* papaya *(Cariea papaya),* etc. But castor and Kesseru are the two important host plants, which are grown in wild or cultivated. Generally, the larva is a tetramoulter (having four moults) and its body grows to considerable extent both in size and weight. In the fifth instar (stage), the larva after maturity (ripe) spins a protective shell of silk (cocoon) around it by secreting fine filament of silk. The larva remains inside the cocoon and metamorphoses into a pupa. The pupal stage is the dormant stage demanding protection. At this stage healthy cocoons (seed cocoons) are kept in grainage for continuance of the life cycle or cocoons are stored in order to allow the emergence of moths before proceeding for extraction of silk.

At the onset of favourable conditions the pupa emerges as the moth (pupal transformation) by piercing the cocoon shell. Moths of opposite sex mate and the mated female moth after separation lays eggs to repeat the life cycle. The number of repetitions of the life cycle depends upon the voltinism (repetition if life cycle), which is proven to be influenced by eco-climatic factors.

Selected, clean and empty eri cocoons collected for spinning are stifled either exposing to sun for one or two hour, or lightly roasted is the steam to fix the silk for preparation of silk yarn. Then they are boiled is a solution of soda of potash or specific leaf or wood ashes to soften the gum and thereby loosen the silk strand. As the eri cocoons are open-mouthed and not composed of continuous filaments they are spinned or spuned by *"takli", "charakha"* or processed on machines in spun silk mills to produce the yarn. The eri yarn so obtained is woven into silk fabrics either with help of loom or machine.

A Silk whose Time has Come

The interest on eri culture has been growing recently. Researchers, scientists, technicians, extension personnel, programme managers and stakeholders spotted out the immense potential for eri silk and its suitability to almost all regions of India.

In the past, the main reason hindering the spread of eri culture from the North-eastern Region (NER) to other states in the country has been the perception that its production elsewhere would be uneconomical. In NER, eri culture is conducted much for food as it is for silk. Besides, almost every housewife there, is an expert spinner and weaver who could convert eri cut cocoons into *chaddars* required by the entire family, thus giving economic sense to eri culture. It was thought that as similar food habits and socio-cultural pattern did not prevail outside the NER, eri culture in other states would not be remunerative.

This view would no longer hold good as the recent advances in eri culture technologies and their dissemination have increased the quality and productivity levels in all stages of production including host plant cultivation with high yielding varieties. Added to this, is the prospect of intercropping of eri host plants with a mix or individually of grams, pulses, other oilseeds and vegetables to increase the unit area returns. The introduction of newer apparatus for spinning of eri cocoons has given rise to finer yarns paving the way to a multiplicity of designs and products including blends. The new technology of slivering eri cocoons for mill processed spun yarn has also opened up several other possibilities in creative weaving.

Currently, the increasing demand for eri silk products outmatches production.

These new developments in the sector coupled with the market generated for eri silk products should do well to draw in many other states to actively promote eri culture in regions where castor, kesseru or tapioca plantations are grown. En silk larvae are handy and can be profitably reared even in hot, humid and rainfed areas providing additional income to poor and marginal farmers on a sustainable basis.

One factor obstructing the work on the eri fibre in design development, product diversification, and launch of products on commercial scales is the non-availability of cocoons or yarn in sufficient quantities in the market. It can be solved by a new package of practices for eri food plant cultivation and eri silkmoth rearing and development of appropriate technologies and machineries for post-cocoon and post-yarn processing.

Market potentiality of eri in recent time has led to its inevitable spread to other parts of India. The fact that *castor* and *tapioca* that can grown under rainfed conditions and on wastelands are the primary food-plants of eri silkmoths could be a boon for the resource of the poor small and marginal farmers in the backward regions of the country. By taking up ericulture in their castor or tapioca plantations, they can expect substantial supplementary income. It is adequately proven by research studies that ericulture would not adversely affect the normal yields of oilseeds in castor or the tubers in tapioca. State Government would have to give special attention to the spread of eri culture. It is a low risk poverty-alleviating vocation in rural and semi-urban areas. Disease incidence and cost of production are much less in ericulture when compared to mulberry sericulture. Presently, many spun silk mills in the country are lying closed for want of silk waste of acceptable quality in adequate quantities. These mills should be able to utilize eri cocoons and produce high quality spun yarn.

Multi-Farious Utility of Eri Products and Bye-Products

Eri Silk

Nowadays, eri silk has a high demand in the national and international market due to its multifarious utility for human happiness and comfort. En silk though lacks the bright lusture, the characteristic of any other silk but has a rugged, smooth with a look of cotton, warmth of wool and ubduded shine of silk. It has better elasticity, higher durability and higher thermal and hygroscopic qualities. It is accepted as an alternative fibre to wool. En fabric is regarded as an ideal fabric for cold countries. Often the fibre spun from eri cocoons is called *"Heavenly thread"* because of its graceful glossy and large diameter, higher elongation rate and stainability.

Eri silk is used for garments, sports-were, bed-sheets and as a blending material for wool. These materials have high demand in the national and international market due to their longer durability. Fri silk also has bright scope for blanket and other winter garments.

Characteristically eri silk is health friendly. Fri silk powder has also been added to food, medicine, cosmetics and soft drinks.

Due to its multifarious utility eri silk has a high demand in the international market. Thus, owing to its overseas demand and export potential ericulture can help to earn foreign currencies, which can aid to national income.

Larvae, Pupae and Moths

The tribal especially in the north-eastern region of the country use the larvae and pupae as food of deliciously. The pupae are highly nutritious with very rich protein content.

The dead larvae, pupae and moths are used as poultry feeds.

Silk Gland

The silk gland of eri larvae is used as surgical tools. Nowadays, Spain is the only country in the world that is producing surgical guts out of the silk gland.

Litters

The litters of eri larvae are the source of organic manure for garden crops and ornamental plants. It can be used as an effective insecticide in paddy fields. The litter can be used for mushroom culture also.

Nothing is Left Unused in Ericulture

In ericulture nothing is left unused as each and every product and bye-product has potential utility for mankind.

Silk, Larvae, Pupae Moth, and Litters

The silk, larvae, pupae moth and litters have high commercial value and menace utility so far economic importance is considered.

The Entire Floss (shell) is utilized

The entire floss (shell) of the eri cocoon is utilized for spinning; even the Pilate is used for making rough yarns which are used as bordering materials for fabrics and shawls.

The Host Plants

Castor, the principal host plant provides dual income to the rearer. The leaves are utilized for rearing of eri larvae where as the castor seeds have high commercial value. The seeds are valuable raw materials for various industries (soap, medicine, lubricating, detergent and coating etc). The tapioca tuber is consumable as food.

The dead and dried host plants are used as firewood suppressing any temptaion for felling of trees for the purpose, a practice at the country side India.

The leaf excess and litters can be made to compost and used as manure for the crop lands.

Ericulture is Eco-friendly

So far the eco-friendly nature of ericulture is considered it reveals that, ericulture does not disturb the existing natural ecosystem. The host plants are mostly nature grown and their cultivation is also simple. This does not require much fertilizer and insecticides.

Ahinsa Silk

In ericulture the pupa is not killed as in other silk culture. As the yarn is spun after the emergence of the moth the eri silk is thus called *Ahinsa silk.*

Ericulture is valuable so far its utility and importance are taken up from the socio-economic angle as well as ecological front.

Ericulture in Orissa

Eri silkmoth *Samia ricini* Donovan is cultivated is limited pockets of rural Orissa. Generally, the rearing is conducted 4-6 times in a year and leaves of the castor plant *(Ricinus communis)* either grown in wild or cultivated are fed. The rearing period spreads throughout the year depending upon the availability of host plants. At present the castor population in the state found fluctuating between 30 and 40 millions and only 4730 rearers are found practicing ericulture. These rearers are mostly belonging to the tribals and economically weaker sections of the society, however, out of these only 400 rearers are being assisted officially. Each rearer produces about 2.5 kg to 4.0 kg of cut cocoons i.e. 15 kg to 17 kg of green cocoons per 100 Dfls where as about 8 kg to 10 kg of cut cocoons are produced by a rearer in Assam. Generally in Orissa rears are found practising ericulture to meet their own domestic demand as a pass-time habit but not in a commercial way to generate income. The rearers also follow a traditional method of culture without research and extension support from the Government or Non-governmental agencies. However, considering the vast natural population of the castor plants, suitable climate and ideal manpower, ericulture in Orissa seems to have bright prospects. Steps are also being taken to proliferate ericultural activities is the state looking at the export potential and overseas demand of eri silk.

TABLE 1
Mean Value of Environmental Parameters of the Study Site during 1997-2001

Study parameters	*Summer (Apr-Jun)*	*Rainy (Jul.-Aug.)*	*Autumn (Sept.-Oct.)*	*Winter (Nov.-Jan.)*	*Spring (Feb.-Mar.)*
Max. Temp. .(°C)	33.01	29.05	29.11	26.73	29.33
	(0.67)	(1.39)	(0.96)	(0.70)	(0.60)
Min. Temp. (°C)	26.74	24.89	24.24	22.86	24.25
	(1.91)	(1.33)	(1.64)	(0.89)	(1.09)
Mean Temp. (°C)	29.87	26.92	26.67	24.79	26.59
	(1.19)	(1.91)	(0.93)	(0.75)	(0.67)
Highest Temp. (°C)	37.50	31.50	31.00	28.75	34.00
	(1.91)	(1.91)	(2.00)	(0.96)	(1.63)
Lowest Temp. (°C)	25.50	24.5	22.00	21.5	22.00
	(5.74)	(3.78)	(1.63)	(1.00)	(0.00)
Duration of sunshine	276.18	160.72	205.22	273.31	270.55
(hr/month)(0.2/0.4 cal/cm^2/min)	(1.56)	(33.35)	(8.57)	(21.36)	(2.69)
Day length (hrs.)	12.54	13.49	11.77	10.55	10.35
	(0.46)	(0.56)	(0.26)	(0.61)	(0.59)
Relative humidity (%)	70.62	82.62	84.77	77.76	73.61
	(6.01)	(3.29)	(4.43)	(5.33)	(8.45)
Total rainfall (mm)	215.95	595.32	426.50	59.25	17.20
	(142.45)	(187.24)	(213.21)	(71.28)	(18.35)
No. of rainy days (≥1	12.25	25.75	20.25	3.50	2.00
mm rainfall)	(4.20)	(0.83)	(1.63)	(1.64)	(1.58)
Total stormy weather	11.75	14.05	11.13	1.19	204
period (hrs)	(4.02)	(4.77)	(3.46)	(0.88)	(1.42)

Note: Figures in the parentheses are SD values.

The commercial productivity of the cultivated Eri silkmoth *Samia ricini* are influenced by a number of factors, viz., nutrition, temperature, humility, photoperiod, rainfall, crowdung etc. In Orissa, the reproductive and commercial productivity are found to vary with variations in the seasonal climatological and environmental parameters of the rearing environment (Tables 1, 2 and 3). The reproductive performances such as, oviposition, fecundity, hatching and incubation period are influenced by the variations in environmental parameters with a mean value of 99.33 ± 0.30%, 328.35+14.96 in number, 91.45 ± 0.92% and 9.54 ± 0.47 days respectively. The fecundity was found to be lowest (309.25 ± 3.02) in summer season and highest in winter season (350.00 ± 14.14). The, seasonal temperature, humility and photo period seems to be influential in affecting the process of oviposition and fecundity as reported by Chowdhury (1970), Jolly *et al.* (1979), Sarkar (1980), Gomma Ahmad (1972) and Begum *et al.* (1995).

TABLE 2

Commercial Traits of Cocoon of Eri Silkmoth *Samia ricini* Donovan in Different Seasons (mean of mean values)

Prevailing season	*Cocoon weight (g)*	*Pupal weight (g)*	*Shell weight (g)*	*Shell ratio (%)*	*Total Silk production (g)*
Summer	1.88	1.61	0.27	14.77	18.94
	(0.10)	(0.06)	(0.02)	(1.90)	(1.78)
Rainy	2.33	1.86	0.46	19.98	33.34
	(0.26)	(0.14)	(0.14)	(5.45)	(10.25)
Autumn	2.48	2.07	0.41	16.76	32.85
	(0.21)	(0.12)	(0.01)	(0.46)	(1.45)
Winter	2.64	1.94	0.71	26.75	58.22
	(0.05)	(0.20)	(0.18)	(6.98)	(15.62)
Spring	2.66	2.13	0.53	19.80	42.49
	(0.09)	(0.08)	(0.11)	(3.77)	(9.64)
Average	2.39	1.92	0.48	19.62	37.16
	(0.32)	(0.21)	(0.16)	(4.72)	(14.47)

(Figures in the parentheses are SD values)

The commercial traits of the cocoon such as cocoon weight, pupal weight, shell weight, shell ratio and total silk production were also found varied significantly during different seasons of the year with an average value of 2.39 ± 0.32 g, 1.92 ± 0.21g, 0.48 ± 0.16 g, 19.62 ± 4.72% and 37.16 ± 9.64 g respectively (Table 2). Higher values of cocoon weight, shell weight, and shell ratio percentage and total silk production were found in winter, spring, autumn and rainy seasons while these values are recorded the lowest during the summer crops (Kar, 2004).

The rearing efficiency of eri silkmoth is found to be the highest (82.17 ± 3.13%) during winter season and the lowest (70.74 ± 2.64%) during summer season. The average ERR percentage is recorded as 76.83 ± 5.20%. The seasonal rearing efficiency seems to follow an increasing trend from summer to winter season, while it follows decreasing trended from spring to summer season. So far commercial productivity is concerned, the highest number of cocoons (26448.24 ± 1232.2 1/100 Dfls) is produced during winter season while the lowest number of cocoon (19943.75 ± 32 16/100 Dfls) is recorded during the summer season. Similarly, the number of green cocoons per kg is found to be the highest (53.19 ± 18.24) in summer season and the lowest (375.94 ± 9.95) during spring season with an average of 423.81 ± 64.14 cocoons per kg. In terms of quality and quantity of silk produced, winter season is found to be the best season of the year. The silk production is found to be the lowest during summer season as cocoon weight and shell weight are recorded lowest during that season. Similarly, while qualitative production by weight is considered, a maximum of 69.82 kg of green cocoon or 18.78 kg of cut cocoon could be produced form 100 Dfls of rearing during winter season followed by the spring, autumn, rainy and summer seasons. On an average 54.44 kg of green cocoons or 10.93 kg of cut cocoons can be produced from each 100 Dfls of rearing (Kar, 2004). However, earlier Chowdhury (1970) has mentioned that approximately 125 kg of green cocoons can be produced from 100 Dfls when 250 cocoons will be obtained per rearing and cocooning ratio will be 95 per cent.

The commercial productivity of ericulture is found to be varied in different seasons of the year depending on the

TABLE 3
Crop Assessment/Cocoon Harvest of Cultivated Eri Silkmoth *Samia ricini* Donvan

Season	*ERR (%)*	*Average weight of cocoon (g)*	*No. of cocoon/kg*	*No. of cocoon/100Dfls*	*Wt. of green cocoon/100Dfls*	*Wt. of cut cocoon/100 (kg)*
Summer	70.74	1.88	531.91	19943.75	37.49	5.38
	(2.64)	(0.10)	(1824)	(3216.78)		
Rainy	71.75	2.33	429.18	21266.75	49.55	10.21
	(3.02)	(0.26)	(21.36)	(481.72)		
Autumn	79.62	2.48	403.22	22261.00	55.21	9.13
	(1.32)	(0.21)	(23.40)	(427.94)		
Winter	82.17	2.64	378.79	26448.25	69.82	18.78
	(3.13)	(0.05)	(6.45)	(1282.41)		
Spring	79.86	2.66	375.94	23975.50	63.78	12.70
	(2.77)	(0.09)	(4.95)	(1653.03)		
Average	76.83	2.39	423.81	22779.05	54.44	10.93
	(5.20)	(0.32)	(64.14)	(2524.32)		

Note: Figure in parentheses are SD values.

seasonal climatic conditions. The winter season seems to be the most favourable one in order to produce both qualitative and quantitative eri silks in Orissa.

Ericulture is a Potential Income Generating Sector

In India En silk historically has been called as the *"silk of the rural people"* or *"poor man's silk"*. Ericulture is also popularly called as *"poor man's friend"* as the rural poor mostly found practicing it for their subsidiary economic gain. In ericulture the silk flows from poor to the rich, but on the other hand, money flows from the rich to the poor.

So far ericulture in Assam is considered nearly 3.5 lakh people belonging to 90,000 families are found engaged to earn their livelihood through ericulture. The total eri silk out put in Assam is also estimated are 950 MT which is about India's total eri silk production (Choaba Singh and Berchamim, 2001).

In Orissa, the poor rural individuals can be recommended to accept ericulture as an occupation for income generation to sustain life, as it assures sure sort return with very low capital investment. Besides, ericulture does not require expertise knowledge and very hard labour, the idle manpower like, older people and house wives can easily share their labour to generate income staying at their own dwelling place. As in Orissa large amount of castor plants are found grown is wild the landless poor sections can easily accept ericulture as an occupation for income generation. However, the rearers can cultivate castor plants in the waste lands or in the croplands to obtain oilseeds as well as to practise ericulture by the use of foliage (40% defoliation of mature leaves does not significantly affect oilseed production).

Castor Agronomy

Castor is generally grown for its oil-yielding seeds. The oil content of the seeds varies from 35-58% in different varieties. Castor-oil is being used for various purposes. The oil is used as a raw material in the preparation of lubricants for high speed engines, preparation of printing inks, transparent paper and soaps etc. It is also used for medicinal and lighting purposes. In eri producing areas leaves are fed to eri silkmoth larvae.

The main castor growing countries are Brazil, India, the USSR and Argentina. India ranks first in area (448,000 ha) but second in seed production (140,000 tonnes), Brazil being the first. Andhra Pradesh (67.2%) Gujarat (12.7%), Karnataka (7.1%) and Orissa (5.8%) account for over 90% of area and also production (Anonymous, 1980).

Castor is available both in the annual and perennial types. The annual type is dwarf and cultivated for seeds. The perennial castor grows wild and is profusely branching and grows almost like a tree, yielding abundance of large seeds and leaves. Nowadays several improved strains of high seed yielding castor varieties have been evolved. But no attempt till now has been made to improve the high leaf yielding variety. However, there are several improved varieties suiting to the different agro-climatic conditions already recommended so far agriculture and seed yield are concerned. The improved varieties cultivated at present are early maturing and take about 150 to 180 days. The castor varieties differ in branching habit of plants, colour of the stem and branches (red and green), the nature of capsule (smooth and spiny), duration of maturity (early or late) and the size of seed.

Castor is tolerant to drought and grows well in relatively dry, warm regions having a well-distributed rainfall of 50 to 75 cms In the heavy rainfall areas, the crop puts an excessive vegetative growth and assumes a perennial habit and last for 3 or 4 years. The perennial variety is resistant and is possibly the indigenous variety. Sometimes the same variety of castor plant behaves both as an annual or a perennial type in the same field.

Growth

Castor grows well on almost all types of soils. It can be cultivated in diverse climates and on poor sandy to rich alluvial soils (Jolly *et al.,* 1979). Some annual and perennial varieties are grown in higher altitudes. Castor requires a moderately high temperature (200-26°C), with low humidity throughout the growing season to produce maximum leaf yields (Anonymous, 1980). Inferior soils, not fit for valuable commercial and food crops, can be used for growing castor.

The crop is sown mostly in June and July and to a limited extent in August and September. It is raised chiefly as a rain-fed annual crop, but sometimes it is planted on the bounds of irrigation channels and borders of garden crops and is allowed to stand for several years as a windbreak or as an ornamental plant.

R. communis is propagated through seeds. Under optimal conditions of heat, light, moisture and aeration, seeds germinate in 7 to 10 days. It, however, takes longer duration to germinate when temperature is low and moisture is less. The cultivated varieties are early maturing and take about 150 to 180 days to be matured. However, the nature grown wild varieties are late maturing, highly branched and produce many leaves and last for three to four years, Leaves of the immature plant are larger in size while leaves of the mature plant are smaller to moderate and the number of leaves are also much more. At about 120 days a perennial variety of castor plant grows up to a height more than 7′ with nearly 8-15 or more branches and about 250-500 leaves. So far leaf yield is concerned, the perennial plants are more suitable as these provide maximum yield. In order to get more branches and leaves, the perennial variety requires light pruning when it attains a height of 5' (Sarkar, 1980). Castor cultivation or planting does not require manure or fertilizer. However, in infertile soils castor cultivation is recommended for increasing productivity.

Castor is grown in many parts of Orissa. It is grown wild as a self-generating crop on hillsides, bands and waste lands. However, it is also cultivated along with other garden crops specifically for getting oilseeds. The leaves of wild castor plants are generally used for ericulture.

Leaf Yield

A mature plant produced 4.38 ± 0.63 kg, 5.4 ± 0.31 kg, 4.20 ± 0.23 kg and 1.78 ± 0.66 kg of castor leaves during the 1st, 2nd, 3rd and 4th year respectively. In toto, a castor plant produced 15.85 ± 0.83 kg of castor leaves in green weight. It is also noticed that as the plant grew in size and age, the number of primary and secondary branches increased, but no primary branch appeared after maturity, and only secondary branches appeared in the subsequent years. But, during the

4th year, the leaf production was very low. Besides, the size and weight of the leaves were found to decrease from the first year. It was also seen that the castor plants show luxuriant vegetative growth during the rainy, autumn and winter seasons. Gradually, the leaf productivity was found to decrease in the spring and summer seasons due to shortage of water in the soil as well as due to increased environmental temperature and decreased humidity content in the air. A systematic castor plantation needs regular manuring and pruning in order to produce maximum leaves (Kar, 2004).

According to standard procedure for castor cultivation (Chowdhury, 1970; Jolly *et al.,* 1979 and Sarkar, 1980) nearly 1300 castor plants can be planted about two meters apart in an acre of land, they can produce 5694 kg of castor leaves in the first year and this quantity of leaves can be plucked at intervals as per requirements for the rearing of eri larvae. Thus, about 20,603 kg of leaves can be produced from one acre of castor plantation during the four years of the crop if environmental conditions are favourable and rainfall is adequate. Out of the total leaf yield 40% i.e. 8,242 kg leaves can be utilized for ericulture as recommended by Mishra (1999) for aruna variety with minimum reduction in castor bean production. But, this needs further investigation, whether it is true in case of local perennial varieties or not, under the environmental conditions of Orissa.

Food Consumption

The total food consumed by a larva during the larval period was 35.15 ± 3.06 g. The quantity of food consumed was progressively increased as the larva passed from first instar to fifth instar. The food consumption showed variation according to the seasonal variations of the climatological parameters. It was found to be the least (31.84 ± 4.08 g.) in the summer season and the highest (39.87 ± 3.96 g.) in the winter season (Kar, 2004).

Economics of Ericulture

It is estimated that, one acre of systematic castor cultivation can provide an income of about Rs.30,000 per annum with an investment of about Rs.3,500 with very little part time labour.

The estimate of income and expenditure is given as under.

Total plantation area	1 acre
No. of castor plants (At about two meters apart)	about 1300
Leaf yield	195,500 kg
Leaf to be utilized for ericulture (i.e. 40% of total yield)	7800 kg
Total Dfls to be reared	900
Total No. green cocoons to be produced	2 lakh
Weight of cut cocoons	100 kg (approximately)
Total oilseed production	350 kg (approximately)

Income

From cut cocoons 100 kg × @ Rs.300/kg (price as in June, 2006 for good quality cocoon)	Rs.30,000.00
From oilseeds *350 kg* × @ *Rs.10/kg*	*Rs. 3,500.00*
Total	Rs.33,500.00

(The price of cut cocoons and oilseed but fluctuates as per production and market demand)

Expenditure

For cultivation	Rs.1,500 (Approximately) (Ploughing, mannuring and labour)
Raring equipments	Rs.2,000
Net gain	Rs.30,000 (excluding gain from eri bye-products and plant bye-products and leaving rearer's labour involved in ericulture).

Thus, owing to the potentiality for income generation ericulture seems to be much profitable and the rural poor can easily practise it for their economic gain.

Advantages that Suit Ericulture in Orissa

The nature of the silkmoth and the easy cultural operations suit ericulture in rural areas.

Nature of the Moth

(i) Erimoth is multivoltine.

(ii) Eri larvae are polyphagus.

(iii) Eri silkmoth is domesticated.

(iv) Eri silkmoths can sustains trough wide fluctuation of temperature and moisture.

(v) Eri moths and larvae are lees susceptible to diseases.

Eri Larval are Domesticated

The eri larval are neither wild as muga and tasar where their fate is entirely depend on nature nor so domesticated as mulberry silk larval, where even slightest negligence may result in heavy crop loss.

Rearing Cost

The rearing cost is just equivalent to the labour used in feeding the eri larvae. Even the rearing equipments are very cheap.

Cocooning

Besides bamboo or plastic cocoonage cocooning is done on dead twigs.

Spinning and Yarn Production

Eri silk spinning is very simple and can be done by cheap appliances like *"takli"* or *"charkha"*. The weight loss during yarn production is very less.

Knowledge of rearing

No expertise knowledge is required for ericulture. Any person with little knowledge can be a successful rearer.

Rate of Rearing

The ERR% of ericulture is very high (nearly 80%) as compared to other silk culture.

Capital Investment

Ericulture need very low capital investment. The poor rural people can conduct ericulture easily with the assistance of government or non-government agencies.

Pricing of Eri Silk

There is not fixed floor price for eri cocoon or yarn. The price fluctuates as per production and requirements. The price of cut cocoons fluctuates between Rs.200 to Rs.350 per kg depending on the market demand.

Ericulture is Profitable

Ericulture provides gainful employment to the tribals and economically weaker sections of the society. Besides monetary gain through silk and oilseed selling a rearer can gain profit from each and every product and bye-products of ericulture.

Prospect of Ericulture in Orissa

Ericulture is simple and does not require any skilled hand at any stage. The tribals and rural poor can accept ericulture as a subsidiary occupation for their economic gain. Hence, both horizontal and vertical development of ericulture is highly necessary for the development of rural economy. This can be achieved through:

1. Large scale rearing of moth.
2. Organised plantation of host plants.
3. Implement of modern methods of spinning.
4. Creating sound marketing facility.
5. Fixing attractive floor price of cocoons and silk.
6. Developing post cocoon technology.
7. Motivating the rearer to conduct commercial large scale rearing.

Conclusion

In view of the existing demand for erisilk in India as well as is the international market, it is high time for Orissa to take necessary steps to develop ericulture in the state in its poverty alevation polices and programmes. Ericulture seems to be the only income-generating sector that can provide gainful employment to nearly 50% of the population of Orissa that are below the poverty line. Ericulture is the only sector, which can provide economic gain to the tribals and poor not only to boost their economy but also to make Orissa rich and prosperous.

REFERENCES

Anonymous (1980) *Handbook of Agriculture*. Indian Council Agricultural Research, New Delhi.

Begum, R., Handique, R. and Hazarika, L.K. (1995) Effect of varying humidities and constant temperature on the development of eriworm, *Philosamia ricini* Boisd. (Lepidoptera : Saturniidae) *Bull. Life Sciences,* 83-88.

Choaba Singh, K. and Berchamim, K.V (2001) Eri: Product diversification pays. *Indian Silk,* 40(1): 11-12.

Chowdhury, S.N. (1970) *Ericulture*, Sericulture Training Institute and Sericultural Research Station, Titabar, Assam, pp. 1-68.

Gomma, Ahmad A. (1972) Biological studies on the eri silkworm, *Attacus ricini* Boisd. (Lepidoptera: Saturniidae). *Ind. J. Seric.,* 11(1): 81-88.

Indian Silk, 40(I): 11-12.

Jolly, M.S., Sen, S.K., Sonwalker, T.N. and Prasad, G.K. (1979) *Non-mulberry Silks*, FAO Manual on Sericulture, Food and Agriculture Organisation of the United Nations, Rome, p. 178.

Kar, J. K. (2004) Studies on Biology, Ecology and Commercial rearing of wild Eri silkmoth Sama ricini Donovan(1798) in Orissa. Ph.D. Thesis (Unpuplished), Utkal University, Orissa.

Sarkar, D.C. (1980) *Ericulture in India*. Central Silk Board, pp. 1-51.

CHAPTER 11

Artificial Recharge of Groundwater for Sustainable Agriculture: A Case Study of Rangailunda Block

Durga Madhab Tripathy*

Introduction

Rangailunda block is situated in the south part of Ganjam district. It consists of 21 numbers of gram panchayats and 82 numbers of villages. The block headquarters is situated at Kanisi. The geographical area of this block is 223.01 square kilometres.

Location and Accessbility

Rangailunda block is featured in Survey of India toposheet Nos. 74A/11, 74A/12, 74A/15, 74A/16, and lies between latitude 18°58´54" N to19°15´55" N and longitude 84°40´07"E to 84°49´55"E. The block headquarters Rangailunda is situated at a distance of 180 km from state capital Bhubaneswar. The block headquarters is connected with gram panchayats and villages by metal and *kutcha* roads.

Population

The total population of the block as per 1991 census is 1,17,277. The scheduled tribe and caste population is 286 and 22,489 respectively. The density of population of Rangailunda block is 525 per square kilometre.

*Groundwater Survey and Investigation Division, Ganjam, Berhampur, Department of Water Resources, Government of Orissa.

Land Use Pattern

The following is the land-use pattern of the block:

i.	Area under forest	0.42 sq. km
ii.	Area under cultivation	151.53 sq. km
iii.	Cultivable Waste	3.96 sq. km
iv.	Area not suitable for cultivation	48.46 sq. km
v.	Area under crystalline	218 sq. km

Need for Groundwater Development

Agriculture is the backbone of the people. Near about 60 per cent of the total population depends on agriculture for their livelihood. Due to erratic and uneven distribution of rainfall, the block suffers from frequent droughts. About 40 per cent of total geographical area of this block is under cultivation of which only 40 per cent is being irrigated in Khariff season and only 15 per cent in Rabi season, mainly from surface water source. The development of groundwater resources in the block is not much.

From the results of groundwater survey and investigation carried out in the block, it is ascertained that there is much scope for development of groundwater resources through installation of suitable structures. Groundwater is most dependable and assured source in the year of low rainfall and dry spell days of monsoon. Optimum utilization of groundwater resources will lead to maximize agricultural production and to improve the socio-economic condition of the people.

Physiography

From the study of the physiography, Rangailunda block can be broadly classified under two physiographic divisions:

(i) The river undulating plains with variable ground elevation from 20 to 80 mtrs above mean sea level.

(ii) The mountainous plateau region in the west with variable altitudes from 100 to 300 m above mean sea level. The eastern ghats run along the block.

The mountainous plateau region is the continuation of the great range of Eastern ghats. The semi-evergreen forests are seen in different parts of the block. The land is gently slopping with appreciable erosional plain having cultivable lands.

Geomorphology

Geomorphologically the block can be broadly divided in two divisions:

(I) River undulating plains and

(II) Hills and mountainous plateau.

Drainage System

The drainage pattern of the area is closely connected with the underlying geological formation and structural features such as folds and lineaments. The rock type and its structures of the block mainly control the drainages of the block. Nalas provide adequate drainage facility for the block. Thus basing on the drainage pattern details, the area can be considered as high groundwater potential zones in alluvial, low to moderate potential in buried pediment and low to poor in high lands.

General Geology

The important rock types are encountered are granites, granite gneisses and Khondalites of the Archean age. The local geology as follows:

Recent	—	Alluvium
Archean	—	Granites and granite gneiss
		Khandalites.

The Granites and Granite gneisses which happen to be the major rock types here occupy the majority of hills and ridges. These quartz, feldspathic rocks are quite resistance to weathering and hence suffered less decomposition and disintegration. But normally they are massive in nature. They are jointed and fractured, the strike joints predominating the dip joints. The presence of garnetiferous Granite gneisses show the metamorphic effects to which the rocks were subjected.

Rock Type

The block is mainly comprising of the igneous and metamorphic rocks belonging to the Easternghat group of Archean geological time. The eastern ghats group is only one type of geological formation in Rangailunda block. The rock types in this group include Khondalite, Charnockite, Granite gneiss etc. These intrusions mainly occur as massive denudational hills, dykes, vein, ridge etc. The alluvium of recent geologic time occurs only along the river.

Soil Type

The important soil types encountered in the area are alluvium, clayey loam and sandy clay. The sandy loam has a thickness of about 2.0 m on an average and being porous and open textured allow the percolation of the surface water through them thereby augmenting the groundwater potential of the area.

Type of Aquifer and its Distribution

The aquifers, which are water bearing zone occurs in unconfined condition above base rock. Different types of aquifer encountered in different village of the block are described as follows.

Weathered Rock

The weathered zone forms the most potential groundwater reservoir at shallow depth. As encountered, the weathering pattern varies from place to place depending upon the topographical set up and rock type. The depth of weathering is more in granite, granite-gneiss and khondalitic formation, occupying undulating plains and topographic depression of pediments. The thickness of weathered zone in granitic terrain usually ranges from 1.5 to 5 metres. In Khondalitic terrain the thickness varies from 2 to 8 metres.

Fractured Zone

The fractured zone mostly occurs below weathered mantle. In the fractured zone under semiconfined to unconfined condition groundwater occurs. Usually one or two fractured zones are encountered within 40 metres below weathered mantle. Beyond this depth occurrence of saturated fracture are uncommon. Water yielding capacity of fractured rock depends largely on origin of fractures (tensile/sheer). The intensity of fracture and extent of fracturing depend upon the nature of hydraulic connection with recharging area. It is also seen that the discharge in the fracture is more at shallow depth or just below weathered residium. The shear fracture yields less water than the tensile fracture. How ever, the occurrence of fracture is not expected every where.

Climate

In Rangailunda block some portion is covered with hills. In hills and platue region the climate is worm and humid. In other portion of block the climate can be treated as moderate.

Rainfall

The normal rainfall of the block is 1295 mm. as per I.M.D records. From the observation of rainfall for consecutive 12 years, the average annual rainfall has been calculated to be 1209.5 mm. The highest rainfall was 1948 mm in the year 1994-95 and the lowest rainfall was 650 mm in 1996-97.Generally monsoon stands from mid June to October. Near about 70 per cent of the annual rainfall is received during monsoon period. In Rangailunda block the behaviour of the monsoon is very peculiar and highly erratic. Sometimes good showers are received in the early part of the monsoon and some time in mid monsoon. Towards the last part of the cropping season rain fall is not uncommon. Besides, the cyclonic rains are also observed towards post monsoon period.

One number of rain gauge station has been fixed at block office. The daily rain fall data has been collected from this rain gauge station.

Temperature

The daily temperature varies from block to block throughout the year. The maximum annual temperature of the block is 36°C and minimum annual temperature has been observed to be 13°C . The atmospheric temperature remains high during last part of may and minimum temperature has been observed during last part of December.

Humidity

The relative humidity of the block varies and remain high during the month of July, August and September and some times in October also. The relative humidity is 63 to 94 per cent during July to September, sometimes in October.

Occurrence, Distribution and Movement of Groundwater

The hydrogeological conditions of the block is mainly controlled by the geological set up, rainfall distribution and the extent of primary and secondary porosities present in different formations for storage and movement of groundwater. Since major portion of the block is underlined by hard crystalline rocks, the geology plays most important role in controlling the occurrence and movement of groundwater resources. The main source of groundwater is the precipitation from rainfall received during the monsoon period and percolated water from irrigated lands, tanks and other water bodies. In the hilly region, much of rain water goes as surface runoff, where as the valley portion and the plain area constitute the main groundwater storage. The hard crystalline rocks do not possess the inter-granular porosity by virtue of their mode of origin. Groundwater occurrence in this formation may only be expected in the weaker foliation planes, fractures, joints and fissures. The presence of joints, fractures activates the process of weathering. The opening along the fracture planes has been improved during the recent geological time due to seepage and chemical action of groundwater. The weathered rocks are the product of excessive weathering of country rock. The weathered mantles are transported from hills or uplands by ephemeral streams and get deposited in low lying area. Just below this unconsolidated weathered mantle, saturated semi-weathered fractured planes are encountered. Under favorable condition groundwater storage in these rocks appears in the weathered and fractured zone.

Water-table Fluctuation

The water-table fluctuations in the shallow aquifers have been observed from three number of monitoring bore wells and one no. of dug well in different topographical situation of the block. The water level observation gives an idea of long-term trend of variation. This parameter has been taken in to consideration for assessing the overall groundwater potential in the area. From the observations it is seen that the average annual fluctuation of water level ranges from 1 to 6 m. It is further seen that the fluctuation is higher in the well located in high land region. The fluctuation has been noticed to be lower in the well of low lying areas.

Groundwater Recharge

The rainfall is the principal source of recharge of groundwater in this area. Near about 70 per cent of the annual rainfall is received during monsoon period i.e. from mid June to October. The source of groundwater is the direct precipitation from rainfall, water from irrigated lands, tanks and other surface bodies like minor irrigation projects and furrows. In the hilly and mountainous region where gradient is high, much of the rain water goes as surface runoff. Hence the groundwater recharge in hilly and mountainous region can be taken as about 5 to 10 per cent. In river plain area, rain fall recharge is higher i.e. 10 to 20 per cent and in flood plain the recharge can be taken as 20 to25 per cent.

Groundwater Quality and its Suitability for Different Use

Four numbers of water sample had been collected from different observation wells. The colour, odour, turbidity, depth of the sample water, temperature, were recorded for physical analysis. The chemical parameters like sodium, potassium, calcium, magnesium, carbonate, bi-carbonate, chloride etc. have been analysed in water quality laboratory with available facilities. The range of water quality parameters is furnishedas from Table 1.

The water samples have been collected from three numbers of observation bore-well (H.P.) and from one number of dug well for base line survey once in a year and trend line survey (limited) three times in a year. Water sample analysis are done in the water quality laboratory located in Groundwater Investigation Division, Berhampur.

It is beyond doubt that the geology and climatic conditions of an area influence the chemical and physical characteristic of groundwater. The minerals content in groundwater is dissolved mainly from parent rock and soil with which the water comes in contact. Therefore, changes in geology is likely to cause variation in quality of groundwater.

Agriculture

The economy of Rangailunda block depends largely on agricultural production. Nearly 62 per cent of the total population depend on agriculture for their livelihood. The principal crops of the block are paddy, ragi, mung, biri, til, groundnut, kulthi, sugarcane and chilly. Usually paddy is sown on rainfed highlands and transplanted in irrigated and medium and low lands. Other crops like *mung*, bin and other grams are grown in large areas of Rangailunda block during rabi season. The oil seeds like groundnut and mustard are also grown both in khariff and rabi season.

TABLE 1
Range of Water Quality Parameters

Sl.No.	Parameter	Unit	Range
1	pH	-	7 - 8
2	EC	m-mhos/cm	372 - 2510
3	Total Dissolved Solids	-	245 - 1605
4	Total Alkalinity	ppm	100 - 260
5	Total Hardness	ppm	70 - 425
6	Calcium (Ca)	ppm	14 - 88
7	Magnesium(Mg)	ppm	7.29 - 76.545
8	Sodium (Na)	ppm	14 – 266
9	Potassium(K)	ppm	0 - 6.5
10	Silica (SiO_2)	ppm	41
11	Chloride (Cl)	ppm	21.27 - 482.12
12	Sulphate (SO_4)	ppm	8 – 69
13	Carbonate (CO_3)	ppm	0
14	Bicarbonate (HCO_3)	ppm	134.2-314.15
15	Total Phosphate (PO_4)	ppm	0

Present Irrigation Facilities

i. **Major Irrigation**: The major irrigation is due to Rushikya canal system and the ayacut area is 12645 ha in khariff.

ii. **Minor Irrigation:** There are 43 nos of minor irrigation projects whose design ayacut 1711 ha in kharif and 196 ha in rabi. 21 numbers projects.

iii. **Lift Irrigation:** There are 9 lift irrigation projects functioning with ayacut are 90 ha in kharif, there is no ayacut area in rabi. The above lift irrigation projects are working under government.

iv. **Water Harvesting Structure:**- There is 9 numbers of Water Harvesting Structure (WHS), maintained by Soil Conservation Department having ayacut 201 ha in kharif. There is four numbers of WHS ayacut area 11 Ha in Rabi. The WHS is located in cultivated land.

v. **Dug well:** In this block dug wells are irrigating 985 ha in kharif. In Rabi 2494 Nos of standard dug wells are irrigating 475 ha. Dug wells are located in cultivated land.

vi. **Other Source:** There are 30 ha and 432 ha irrigated in kharif and rabi respectively by other sources in this block.

Assessment of Groundwater Resources and Potential

Since the major portion of the block comes under hard rock terrain, groundwater occurring in the weathered residium as well as fracture zone largely depends on the thickness of weathered mantle. Rainfall infiltration in the area is the main source of groundwater recharge. Other water bodies like tanks, reservoir, irrigation canals have also contribution for groundwater recharge. The behaviour of water-table in the block has been monitored from three numbers of monitoring bore-well and one number of dug-well. The block-wise groundwater potential available in Rangailunda block has been assessed on the basis of latest norms circulated by Governmrnt of India.

Scope for Groundwater Development

The utilizable groundwater resource of the block has been estimated to be in the order of 5605 hectare meter. The nct annual draft by the existing groundwater structure has been calculated to be 829 hectare metre. Thus the balance groundwater potential of 4378 hectare meter can support installation of further 1700 nos of standard irrigation dug wells with pump sets. The present stage of groundwater development of this block is 20 per cent. Hence there is a vast scope for further development which can bring improvement in the socio-economic condition of the people.

Conclusion

Rangailunda block spans a total geographical area of about 223.01 sq km in Ganjam district with a population of about one lakh 17 thousands. Kanisi is the headquarters of the block.

The southwest monsoon is the primary source of groundwater recharge. The return seepage from surface water bodies like tanks, water-harvesting structures, unlined canals, and irrigational ayacut are secondary sources of groundwater recharge.

The net annual groundwater availability is estimated to be about 5605 HM. The net annual groundwater draft for all uses (Domestic, Industrial and Irrigation) is about 829 HM during 2001. The annual groundwater allocation for domestic and industrial water supplies up to next 25 years is about 398 HM. The net annual groundwater allocated for future irrigation use is estimated to be about 4378 HM.

For sustainable groundwater development, the cultivated area of the block has the potential to support additional 1700 nos. of standard energised irrigation-dug wells. It is advisable to install deep-borewell after conducting VES tests since borewells are not feasible everywhere and are site specific.

The present stage of groundwater development is only 20 per cent in area, which shows that the demand for irrigation. The stage of groundwater development for total block is 20 per cent; therefore the status of groundwater resource is SAFE.

Recommendations

As already discussed, there is enough scope for development of groundwater through suitable abstraction structures e,g dugwell and substantial area can be brought under irrigation for increasing agriculture production of the block. For large scale homogeneous development and effective use of groundwater resources, the following suggestions are made:

- The spacing between two consecutive dug wells and bore wells should not be closer than 150 and 250 metres respectively to avoid interference.
- For optimal utilisation of irrigation potential emphasis should be given for installation of pumpsets in wells, which are being operated manually.

- Revitalisation of old and defunct wells may be taken up by deepening the entire thickness of weathered zone to increase the overall yield of well.
- In areas of declining trend of groundwater, the artificial recharge to groundwater aims at augmentation of groundwater resource by modifying the natural movement of surface water utilising suitable civil constructions technique may be practised in the block. Checkdams, gully plug, contour trenching, subsurface dyke, recharge pit, recharge dugwell are some of the suitable structures required for recharging groundwater reservoirs in the block.
- Maximum irrigation potential can only be created by conjunctive use of surface and groundwater i.e. in keeping a balance between simultaneous development of surface and groundwater structures. This will serve the dual purposes of preventing water logging situation and supplementing the irrigation water requirement particularly at tail ends of canals.
- Optimal utilisation of irrigation potential can be achieved by latest farming and effective water management technique; economic distribution of water by maintaining minimum pumping hours and selecting most suitable cost effective cropping pattern

Acknowledgement

1. Deputy Director Agriculture, Berhampur.
2. Soil Conservation, Berhampur.
3. Deputy Director, Geology and Mining, Berhampur
4. Office of Superintending Engineer Irrigation, Berhampur.
5. Office of Superintending Engineer, OLIC, Berhampur
6. District Statistical Office, Ganjam, Berhampur.
7. Emergency Section, RDC Office, Berhampur.

CHAPTER 12

Agriculture: A Commitment to Society

Subhasmita Mahapatra*

The little word having vast meaning and application to maintain the back-bone of the society is agriculture.

The most important factor in any planning for development, and economic upliftment is that of turning a hungry, discontented people into a happy, well-fed one. Food can be had either by import or by produce its own. The problem, therefore, reduces itself to one of agricultural improvement. A group of people to be involved to make some production with some devices and techniques to produce food and economical bi-products, i.e. through agriculture.

In many countries, most of the people depend upon this god-gift. Hence, there is always a need to increase agricultural productivity. Agricultural development depends upon institutional and technological factors.

The first step to be taken is to increase the fertility of the soil, to ensure a higher yielding of crops per acre. Old farmers relied upon easily available manure like cowdung. Every farmer must be thought to realize that chemical fertilizers can improve the productivity of land to an undreamt to extent. The proportion of organic and inorganic manure depends upon the nature of the soil.

The next step is to improve the quality of the seed. The supply of good seeds is by far the most vital problem. Plant breeding is not only an art, it is a highly specialized science.

*KSVB College, Bhanjanagar.

Fortunately at many agricultural farms, especially in Hyderabad, this is being done successfully. The scientists will help us to raise needs fit for a local environment. It is reported that in former Soviet Union hybridization method was adopted with great success.

Improved method of cultivation must also be introduced. Japanese method of rice cultivation has yielded splendid results, whenever applied. In Europe, the production per acre is much higher than other countries. So high yielding stairs should be used and double harvesting is to be ensured every year. If we want to increase our food supply, not only should production be improved but wastage also must be eliminated by improved methods of farming and preserving surplus food-grains.

There was a time when agriculture in the country was largely a gamble on monsoon, when our farmers had to depend almost wholly on rain water for irrigation, has bullock-driven plough for tilling the soil and an organic fertilizers like cowdung and garbage for manuring the soil. Modern agriculture in the western countries is highly mechanized, and if we are to triumph over the vagaries of nature, we have to supply the farmer with all the mechanical means of cultivation.

In order that we may ensure a rapid transition from primitive to modern agriculture certain basic industries have to be developed on a large priority based.

The high-level agricultural production western countries reached, has come only after the industrial revolution of the past century. Formerly, it was believed that the rate of population growth will for exceed the agriculture has disproved this theory. The transformation of primitive agriculture into a scientific agriculture is therefore, achieved at high cost, entailing enormous sacrifice on the part of the common man.

We are now on the threshold of an agricultural revolution. The major problem now is the preservation of the environment. The sudden rise on the price of crude oil has hit our fertilizer industry. The enormous cost of all these is proving a well high burden. We have got to together own belt and fate the challenge. There is no doubt that once own industrial complexes gets going, the road to prosperity will be fairly smooth and agricultural and industrial progress will supplement each other.

India is largely an agricultural country. Eighty per cent of her vast population live, directly or indirectly, on income derived from agriculture. But to say that India lives by agriculture does not mean that India is agriculturally advanced. Rather she is extremely backward in this respect. Her annual yield of crops per acre is lamentably below what it ought to be.

There was a time when small plots of land, cultivated by individual farmers who followed age-old methods, summed up the position of Indian agricultural system.

The employment of scientific technique, especially in America and Russia, has achieved tremendous progress. In the first place, machines have superseded manual labour. That means a larger average is brought under cultivation in more efficient manner. Secondly, the fertility of the soil has been increased by scientific, i.e., chemical manuring. Thirdly new crops of a better quality and higher yield have been introduced. Fourthly, highly improved methods of irrigation and crop-protection and stoning have been adopted.

Fortunately, India in recent times has brought about a green revolution, resulting not only in self sufficiency but also surplus production of food. India has now a huge buffer stock of foodstuffs and she is in a position to export food grains.

It is necessary to modernize the antiquated outlook of our peasants.

All this needs fan-sighted planning. When the ownership of land is restored to the cultivators and determined efforts are mode to modernize their outlook, agriculture in India will flourish much more.

It is a good sign that the claims of agriculture have been taken into adequate consideration in our successive five-year plans. Now greater emphasis has been placed on the development of agriculture than ever before and it's a good commitment.

Index